Nut Crops - New Insights

Edited by Muhammad Akram

Published in London, United Kingdom

Nut Crops - New Insights
http://dx.doi.org/10.5772/intechopen.102145
Edited by Muhammad Akram

Contributors
Federico Hahn Schlam, Fermín Martínez Solís, Adriana María Descalzo, Sergio Aníbal Rizzo, Carolina Daiana Pérez, Andrea Biolatto, Enrique Alberto Frusso, Gabriela María Grigioni, Luciana Rossetti, Diannita Harahap, Paulina G. Flores, Sunita Mishra, Gunjan Gupta, M. A. Firdaus, Giorgio Calabrese, Cinzia Myriam Calabrese, Muhammad Akram, Sabira Sultana, Nosheen Aslam, Naheed Akhter, Abdul Wadood Chishti, Muhammad Riaz, Walaa Fikry Elbossaty, Rida Zainab, Umme Laila

First published in London, United Kingdom, 2023 by IntechOpen
IntechOpen is the global imprint of INTECHOPEN LIMITED, registered in England and Wales, registration number: 11086078, 5 Princes Gate Court, London, SW7 2QJ, United Kingdom

British Library Cataloguing-in-Publication Data
A catalogue record for this book is available from the British Library

Additional hard and PDF copies can be obtained from orders@intechopen.com

Nut Crops - New Insights
Edited by Muhammad Akram
p. cm.
Print ISBN 978-1-80356-632-0
Online ISBN 978-1-80356-633-7
eBook (PDF) ISBN 978-1-80356-634-4

Meet the editor

Dr. Muhammad Akram is an associate professor in the Department of Eastern Medicine at the Government College University Faisalabad in Pakistan. He obtained his Ph.D. from Hamdard University, Pakistan, in 2013. From 2015 to 2017, he served as chairman of the Department of Eastern Medicine and Surgery, at the University of Poonch Rawalakot in Pakistan. From 2018 to 2022, he was an associate professor and chairman in the Department of Eastern Medicine, Faculty of Medical Sciences, Government College University Faisalabad, Pakistan. Dr. Akram has received many honors and awards during his career. He has won three international projects. He serves as an editor and reviewer of several national and international journals and has numerous publications and presentations to his credit. He is an active member of several professional societies. Dr. Akram's research interests include hyperuricemia, xanthine oxidase inhibition by medicinal plants, enzyme inhibition, Indusyunic Medicine, phytochemistry, poisonous plants, bioactivity and phytopharmaceutical evaluation of herb drugs and their natural products, medicinal plants, biochemistry, and bioinformatics.

Contents

Preface

This edited volume is a collection of reviewed and relevant research chapters concerning the latest developments in nut crops. It includes scholarly contributions written by horticulture experts and reviewed by experts in the biological sciences. Chapters address such topics as nut crop cultivation, dietotherapy with nuts, and the nutritional composition and health-promoting effects of nuts.

Dr. Muhammad Akram
Associate Professor,
Department of Eastern Medicine,
Government College University Faisalabad,
Faisalabad, Pakistan

Chapter 1

The Beneficial Health Effects of Nuts in the Diet

Sabira Sultana, Muhammad Akram, Nosheen Aslam, Naheed Akhter, Abdul Wadood Chishti, Muhammad Riaz, Walaa Fikry Elbossaty, Rida Zainab and Umme Laila

Abstract

Nuts are nutritional containing foods that nature has gifted with complex matrix of bioactives and beneficial nutrients including poly- and monounsaturated fatty acids, antioxidant phenolics, phytosterols, tocopherols, minerals, fibers, and high-quality proteins. Because of their unique nutritional composition, they are beneficial to human health. Many epidemiological studies reported the beneficial effect of nuts on coronary heart diseases as well as in diabetes for both males and females. Nut crops also have good effects on inflammation, vascular damage, oxidative stress, cancer, and hypertension. Interventional studies reported that nut consumption has lipid lowering effect. Nuts are thus among the most highly nutritious foods available. When left in their shells, many nuts have a long lifespan and can be easily kept for winter use. Much research has been conducted on nuts and their beneficial effects on health. This chapter discusses the current knowledge of nuts crops and their beneficial effects on health. Currently available literature suggests that consumption of nuts particularly as part of healthy diet is associated with beneficial outcomes such as better cognitive function, decreased risk of cardiovascular and metabolic abnormalities, and cancer. Further, long-term and interventional studies are required to make a definitive conclusion on the health-promoting effects of nuts consumption through the diet.

Keywords: crops nut, beneficial effects, peanut, walnut, nutrition

1. Introduction

Nuts are a rich source of protein, healthy fats, fiber, vitamins, and minerals. In addition, they contain an amount of unsaturated fats that reduce heart disease. Nuts are difficult to characterize in a way to be acceptable by the botanists and compatible for general use. The groundnut or peanut, for example, is a legume, whereas the chufa nut, which is found in southern Europe and Africa, is a tuber. In some languages, such as French, don't even have a generic term for nuts. In French, "Noix" appears like one, but it simply means walnuts. Nuts consist of dry fruits consisting of one or two edible kernels inside a hard shell. They are included Brazil nuts; cashew nuts; hazelnuts; macadamias; pecans; pine nuts; pistachios; Walnuts; peanuts. They are a rich source of protein, carbohydrates, vitamins (B1, B2, B3 and vitamin E), unsaturated fats, and minerals such as (magnesium, potassium, calcium, plant iron, and zinc), so nuts have a medical importance as it works to maintain weight, diabetes control, heart disease prevention through reducing low-density lipoprotein (LDL). Nuts are extremely healthy and important to people in many parts of Asia and Africa. The majority of nuts are high in fat (e.g., pecan 70%, macadamia nut 66%, Brazil nut 65%, walnut 60%, almonds 55%, and peanut butter 55%). Only very few have a very high content of starch [1], while most have a significant protein level (in the 10–30% range). Many nuts have recently been discovered to be particularly high in antioxidants [2]. Nuts are thus among the most highly nutritious foods available. When left in their shells, many nuts have a long lifespan and can be easily kept for winter usage. Much research has been carried out on nuts and their beneficial health outcomes. The published report from the pioneering Adventist Health Study showed a relationship of nut consumption and lower risk of coronary heart disease (CHD) in 1992 [3]. Ros [4] published a clinical study indicating that a walnut-enriched diet lowers serum cholesterol levels when compared to a balanced diet. A recently published book [3] and the papers of a Symposium on Nuts and Health held at the U.S. Department of Agriculture Studies Laboratory in the University of California at Davis [5] provided extensive information on research published about nuts and their beneficial health outcomes up to 2005. Nuts contain one seed, they are dry fruit, and upon maturity the ovary wall becomes hard. The most consumable nuts are *Juglans regia* (walnut), *Corylus avellana* (hazelnut), *Pistachia vera* (pistachios), *Anacardium occidentale* (cashew nut), and *Prunus amigdalis* (almond). Other commonly consumed nuts are (*Macadamia integrifolia*) macadamias, (*Pinus pinea*) pine nuts, (*Carya illinoiensis*) pecans, and (*Bertholletia excelsa*) Brazil nuts. Pea nut (*Arachis hypogea*) is a legume or ground nut and also comes under the definition of nut by consumers. Peanut contains similar nutrients as other nuts [6]. Nuts, pulses, and seeds are all high in nutrients and have been part of mankind's diet from pre-agricultural era [7]. Nuts are ingested whole (roasted or fresh), in spreads (almond paste, peanut butter), as oils, sauces, mixed dishes, pastries, ice creams, and baked goods in Western countries as desserts, snacks, or as part of meal. Except for vegetarians and other health-conscious community, such as Seventh-Day Adventists, nuts consumption is reduced in most industrialized countries throughout the last century, becoming only a minor source of energy in daily diets [8]. Recently, nut consumption increased in Western region. This food has wide media coverage and also included in healthy eating as these have many beneficial health effects. The nuts are considered as part of optimal diet and are found involved in preventing the chronic heart disease [9]. In 2003, the US Food and Drug Administration reported that intake of nut reduces the risk for CHD and lowers the serum cholesterol level. Since then, nuts have become part of healthy diet [10], and the American Heart Association reported that the nut consumption through diet is ideal for cardiovascular health and overall general health 2020 [11].

The scientific data for nuts as cardio-protective foods come from a combination of epidemiological observations suggesting a consistent inverse relationship between nut consumption and the development of CHD [12] and the several clinical trials revealed the potential benefits of nut consumption on the lipid profile [13] and other intermediate markers of CHD [13, 14]. The synergistic effect of various bioactive elements of nuts, which may all favorably influence the human physiology, is most likely the mechanism for these beneficial effects. As a result, nuts are high in vegetable protein and fat, primarily unsaturated fatty acids. They are also high in dietary fiber [14], vitamins (such as niacin, folic acid, vitamin B6, and tocopherols), minerals (calcium, potassium, and magnesium), and many other bioactive elements such as phytosterols and phenolic compounds. Despite the fact that nuts have a high energy density, evidence from epidemiological research and clinical studies suggests that frequent consumption does not lead to obesity or raise risk of diabetes [15]. This chapter discusses the common edible nuts, their nutrient composition and beneficial health effect.

1.1 Almonds

Almonds are scientifically called *Prunus dulcis* and are also known as sweet almonds. It belongs to Rosacea family. It is a good source of nutrients and is popular in general population as a healthy food. The demands for almonds consumption increases day by day. It consists of many nutritious constituents such as amino acids, lipids, fatty acids, carbohydrates, proteins, vitamins, and secondary metabolites. The environmental and genetic factors affect the nutritional composition of almonds. Almond has therapeutic and protective effect as well. Research confirmed the prebiotic potential of *Prunus dulcis*. Clinical studies reported the modulatory effect of almond on serum lipid, glucose, and uric acid level and its role on body weight, obesity, cardiovascular diseases through raising the levels of good cholesterol and reducing the levels of bad cholesterol, diabetes, and metabolic syndrome [16].

Almond is native to Central Asia, but now it grows all over the world in temperate region [17].

In present days, USA produce almond largely and followed by Australia and Spain. The different varieties of almond have different chemical composition because of different genetic, ecological, and processing conditions. Regular intake of nut produces beneficial health effects against many ailments including cardio-metabolic disease [18]. Several epidemiological and clinical studies reported the beneficial health effect of almond against number of diseases such as metabolic syndrome, hypertension, obesity, and diabetes mellitus [19]. Regular consumption of nuts reduces body weight and improves the metabolic health [19].

The prunus kernel consists of a seed, which is edible part. The seed is covered by brown skin. Externally hull is present, which protects the seed. The hull is an intermediate shell [16]. When fruit mature, hull opens and seed separates. Almonds contain lipids, proteins, and carbohydrates. They have low moisture content and less bioactive compounds. The macro and micro-nutrients present in almonds are responsible for its beneficial health effects [20]. The diversity is present among the verities. The method of cultivation and climatic factors affect the chemical composition of almond [21].

Chemical composition also changes in raw and roasted form. Roasting processes alter the lipid composition and further involve in oxidation process, which lessens the antioxidant compounds. Almond acts as prebiotics that promote gut microbes. The polyphenols and fiber contents act as a substrate for microorganism for microbial fermentation [22].

1.2 Pistachios

Pistacia vera L., commonly known as pistachio, belongs to Anacardiaceae family. It is an ancient nut tree. It is distributed in central and south-west Asia and in the Mediterranean region. Valle del Platani and Bronte are two main cultivars of Pistachio. The soil composition and climate conditions of Asia and Mediterranean regions are much favorable, and the pistachio nuts grown in these regions are of dark green color, good nutritional contents, and pleasant aromatic taste. Its bioactive phytoconstituents, fatty acid composition, and pharmacological activities have been evaluated. Research showed its biological and pharmacological properties [23]. Pistachio contains dietary fibers, protein, vitamins, magnesium, unsaturated fatty acids, polyunsaturated fatty acids, vitamin K, Vitamin B, and vitamin A. It also contains phytosterols, lutein, γ-tocopherol, and polyphenols. All these phytoconstituents are antioxidant in nature. The pistachio nuts have therapeutic role in type II diabetes mellitus, obesity, and cardiovascular diseases. It has antioxidant and anti-inflammatory activities. Regular intake of pistachio nut improves health and dysmetabolic condition and helps in preventing chronic degenerative diseases [24].

1.3 Walnuts

United States produced 504,000-ton walnut in 2010 positioned after almond, i.e., 820,000 ton. *Juglans regia* L. is commonly known as Walnut OR English walnut. Its origin is Persia and is distributed to United States. California produces 90% of English walnuts. Black nut is another variety produced largely by United States. Walnut is largely consumed worldwide. In the USA, almonds, pistachios, and pecans are largely consumed nuts [4]. English and black walnuts are used in cooking items for cereals, candies, snacks, and backed goods. Black walnut has small kernels and thick hard shell, the commercial value of which is lower than that of English walnut.

The clinical and epidemiological studies showed the therapeutic and beneficial effects against several diseases, including diabetes mellitus, oxidative stress, cardiovascular diseases, inflammation, cancer, and neurodegenerative disorders [25]. The beneficial effects of health have been due to the presence of bioactive nutrients such as tocopherls, phytosterols, proteins, fibers, minerals, and fatty acids [26].

In spite of the black nut being rich in a variety of chemical constituents, the English walnut is more investigated.

The importance of nuts as antioxidants is due to the fact that they contain polyphenols compounds that reduce oxidative stress by neutralizing free radicals [27]. A scientific study has proven that walnuts have a greater antioxidant capacity than fish, in addition to protecting the fats in sensitive cells from being damaged as a result of exposure to oxidative stress [28].

A study conducted on 13 people who ate a regular meal of walnuts and almonds confirmed that polyphenol levels increased and oxidative stress was significantly reduced [29]. Research has also proven that eating nuts reduces bad cholesterol levels by 26–33% within 2–8 hours of eating them, and it is known that an increase leads to heart complications [30].

1.4 Cashew nut: *Anacardium occidentale*

Cahew nut botanically known as *Anacardium occidentale* belongs to Anacardiaceae family. Anacardiaceae family has 400–600 species. Among these species, cashew from Anacardium genus is rich source of nutritive constituents. It is a tropical plant distributed in South America and Brazil.

Cashew plant is 5–14 m in height. It has short trunk and irregular branches scattered on the ground. Leaves are green, and flowers are short. Cashew nut fruit is edible and is used for medicinal and food purposes. Cashew nut gained fame in Second World War due to its by-product nut shell liquid.

It has nutritional properties and pleasant taste. Cashew nut lowers the serum low density lipoprotein and cholesterol level, hence preventing coronary disease. Cashew contains fats and protein. Protein includes cysteine, lysine, valine, arginine, and vitamins such as E, D, and C. Cashew gum is used against many ailments. Cashew nut contains many unsaturated fatty acids and less saturated fatty acids. It has health effects against many diseases such as oxidative stress, diabetes, and cardiovascular disease. It is used in diabetes, obesity, heart disease, digestive disorders, urinary disorders, and in several other ailments such as bone relaxation, common cold and flu, etc. It is also effective in cancer and has antiaging properties [31].

1.5 Peanuts

Arachis hypogaea (peanut) belongs to the family leguminoseae. It is also known as ground nut. It is distributed worldwide. Asia is the largest producer of peanuts and accounts for 60% of world production [32]. There are many cultivars of peanuts but four major cultivars include Virginia, Spanish, Runners, and Valencia [33]. Peanut is the rich source of vitamins, fibers, proteins, oils, and carbohydrates. Peanuts contain vitamins, proteins, fibers, oils, and carbohydrates in abundant amount and consumed worldwide due to easy availability and low cost as compared to other nuts [34]. Peanuts are commonly used in snacks, butter, and confectionary. Peanut oil is extracted and used worldwide. In the United States, peanut crop is used in making peanut butter and snacks.

The oil extracted from peanut is rich source of protein, vitamins, minerals, antioxidant, and dietary fiber. The oil is consumed by human beings and used in animal feed. The clinical studies reported that peanut intake reduced certain chronic ailments such as diabetes, cholelithiasis, and coronary diseases in both males and females. The peanut reduces the body weight and also cures hypertension. It has anti-inflammatory and anticancer activity. Peanuts have many health benefits, which are because of their valuable nutrient components. The vitamin B, E, and fibers play protective role in health. Peanuts have economically low cost, easily available and affordable to general population. It can also be used against malnutrition in developing countries due to its nutrient contents. The Indian raw peanut kernel contains 47.27% fat, 17.43% carbohydrate, 5.25% moisture, 25.48% protein, and 1.84% ash. Peanuts contain essential amino acids and are easily digestible. Vegetarians who are not taking animal meal depend upon plant sources for protein such as soybeans and peanuts.

Peanut protein is an alternate source of animal protein. All 20 amino acids are present in peanuts in different ratios. The arginine is present in highest amount. It also contains cysteine and methionine in limited amount [35].

1.6 Betel nut

Areca catechu is the source of betel nut. It belongs to family Palmae. Betel nut is masticatory, stimulant, astringent, sialagogue, sweeten breath, emmenagouge, anthelmintic, resolving, and repercussive. It stimulates the peristaltic movement of intestine. Due to its astringent properties, it constricts bronchial muscles. It is anthelmintic and works against tape worm, round worm and also used in veterinary practice. Beetle nut is also effective in dysentery and diarrhea. This nut is effective in dyspepsia and has emmenagouge properties. Betel nut is astringent used in gingivitis to stop bleeding from gums and gives strength to teeth. It is used in female vaginal discharges and in heart burning condition of pregnancy. Chief use of betel nut in the

Orient is as breath-sweatening masticatory enjoyed by people from centuries and used in Pan. In conjunctivitis and epiphora, the nut is made into ash, made more fine, and applied as collyrium giving relief [36].

1.7 Nutrient composition of nuts

Nuts are nutrient-rich foods. Nuts are the richest source of fat after vegetable oils. They are nutrient-containing plants, **Table 1**.

Nuts	Energy (kJ)	Fat (g)	Saturated fatty acid (g)	Mono saturated fatty acids (g)	Polysaturated fatty acids (g)	Linoleic acid (g)	α-linolenic acid (g)
Almonds	2418	50.6	3.9	32.2	12.2	12.2	0.00
Brazil nuts (dried)	2743	66.4	15.1	24.5	20.6	20.5	0.05
Cashews	2314	46.4	9.2	27.3	7.8	7.7	0.15
Hazelnuts	2629	60.8	4.5	45.7	7.9	7.8	0.09
Pistachios	2332	44.4	5.4	23.3	13.5	13.2	0.25
Peanuts	2220	49.2	6.8	24.4	15.6	15.6	0.00
Walnuts	2738	65.2	6.1	8.9	47.2	38.1	9.08

Table 1.
Average nutrient composition of nuts (per 100 g).

Nutrients	Almonds	Walnuts	Cashews	Pistachio	Hazelnuts	Brazil-nuts	Chest- nuts
Water (g)	4.14	4.07	5.20	4.37	5.31	3.42	48.65
Energy (kcal)	579	654	553	560	628	659	213
Protein (g)	21.15	15.23	18.22	20.16	14.95	14.32	2.42
Carbohydrates (g)	21.55	13.71	30.19	27.17	16.70	11.74	45.54
Dietary fiber (g)	12.5	6.7	3.3	10.6	9.7	7.5	8.1
Total sugars (g)	4.35	2.61	5.91	7.66	4.34	2.33	-
Total lipids (fat) (g)	49.93	65.21	43.85	45.32	60.75	67.10	2.26
Phosphorus (mg)	481	346	593	490	290	725	93
Potassium (mg)	733	441	660	1025	680	659	518
Calcium (mg)	269	98	37	105	114	160	27
Magnesium (mg)	270	158	292	121	163	376	32
Iron (mg)	3.71	2.91	6.68	3.92	4.70	2.43	1.01
Zinc (mg)	3.12	3.09	5.78	2.20	2.45	4.06	0.52
Manganese (mg)	2.179	3.414	1.655	1.200	6.175	1.223	0.952
Copper (mg)	1.031	1.586	2.195	1.300	1.725	1.743	0.447

Table 2.
Nutritive values of nuts per 100 gram.

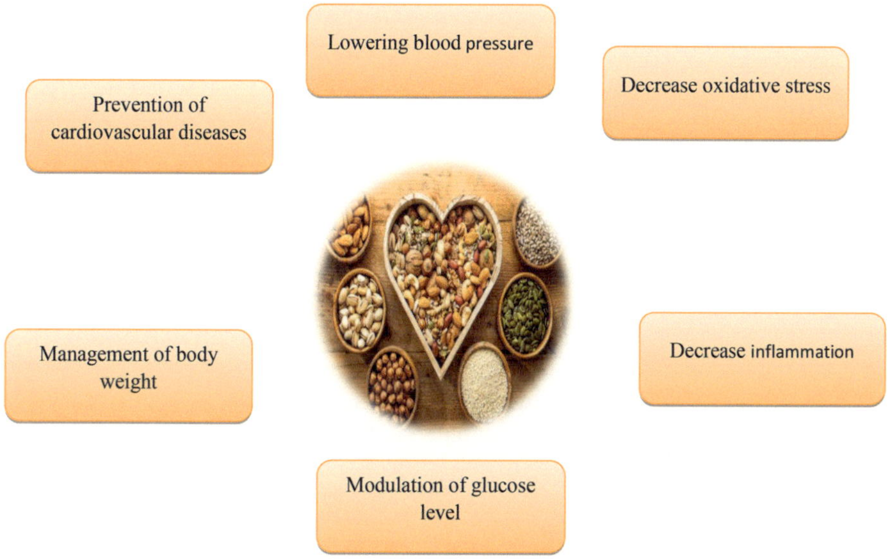

Figure 1.
Medical benefits of nuts.

Fatty acids present in nuts are beneficial because saturated fatty acids present in lower amount. They are also the rich source of bioactive constituents such as proteins, minerals, fibers, tocopherols, and phytosterols, **Table 2**.

1.8 Nuts and health

This chapter was planned with the aim to explore and highlight the beneficial health outcomes identified through nut consumption, which are categorized as oxidative stress, inflammation, prevention of cardiovascular diseases, management of body weight, and the modulation of glucose level in the body, **Figure 1**.

2. Oxidative stress and inflammation

Nuts are sources of phenolic compounds and tocopherols with anti-inflammatory and antioxidant activities. The beneficial effects of cashew nuts on oxidative stress and acute and chronic inflammatory processes have recently been investigated in various in vivo experimental models on mice, such as ischemic injury, lung injury, pancreatic injury induced by cerulean, paw edema induced through carrageenan, osteoarthritis induced through monosodium iodoacetate (MIA), and colitis induced by dinitrobenzene sulfonic acid (DNBS) [37]. Antioxidative, anti-inflammatory, and analgesic activities of cashew nuts have been highlighted and hypothesized that these beneficial effects have correlation with the high phenolic contents in cashews mediating the 5-LOX COX pathway activation [37]. The effectiveness of Brazil nut and pistachio against inflammation and oxidative stress has also been demonstrated through *in vivo* studies [38].

Published study findings suggest that intake of nuts may contribute in the protection against inflammatory processes and oxidative stress, which is mainly because of the abundantly present bioactive metabolites such as carotenoids, flavonoids, and phenolic compounds found in the kernel and skin as reported in published studies.

Clinical and epidemiological studies suggested that antioxidant vitamins, ω-3 PUFA, magnesium, L-arginine, and dietary fiber may play an essential role in the modulation of inflammation. The relationship observed between inflammatory markers and frequent nut consumption has reported mixed results when investigated in controlled feeding intervention [39].

3. Prevention of cardiovascular diseases

Association between nut consumption and decreased risk of ischemic heart disease (IHD), cardiovascular disease (CVD), and the incidence and mortality of CVD has been consistently proven by the epidemiological studies [40]. Consumption of nuts also increases longevity in addition to protection against CVD [41]. Nuts being the dense source of energy due to the high content of bioactives play significant role in decreasing the risk of CVD through multiple mechanisms: by having positive influence on lipid and/or glucose homeostasis [42], hypercholesterolemia [43], obesity [44], oxidative and inflammatory biomarkers, and antioxidant defenses.

Nuts consumption has positive effects on various CVD risk factors including improvements in lipoprotein cholesterol, total cholesterol (TC), and triglycerides [45]. The authors of the published studied attributed that cardioprotective potential of almonds is due to their dense bioactive contents such as phytosterols, dietary fiber, and high poly-unsaturated fatty acids contribute to their conserving HDL-C and LDL-C lowering effects [46]. Later on, it was reported that almonds incorporated [47]. Additionally, walnut supplementation increases the HDL-C after treatment for 6 months possibly due to the higher intake of PUFA.

The reduced risk of CVD through almond snacking by increasing variation in heart rate in healthy adults during mental stress and the consumption of whole almond was linked with better quality diet and reduced CVD risks in the adult population of UK have been demonstrated in a randomized controlled trial study [48]. The authors also reported the association of almond snacks consumption with significant endothelial function improvement, in addition to the lowering LDL-C among adults with above average risk of CVD.

Numerous studies have indicated that eating nuts reduces the risk of heart disease and stroke, due to the fact that they contain polyphenol compounds that reduce the level of bad cholesterol and work to reduce the rate of inflammation in the body by reducing inflammatory markers C-reactive protein (CRP) and interleukin 6 (IL-6) [49].

Also, in one study of five people who took cashews, lower blood pressure was observed in addition to lower triglyceride levels [50].

4. Modulation of glucose

The beneficial role of nuts in the metabolism of glucose and insulin has been associated with secondary active metabolite, minerals, fats, and fiber contents. It has been shown that PUFA, carotenoids such as β-carotene and lutein, γ-tocopherols, and procyanidins content of pistachio [40], PUFA content of walnuts and the fat and fiber content of almonds are responsible for these beneficial effects [44]. A published study reported that walnuts induced significantly positive effect at the initial phases of nutritional change in a PUFA-delivered healthy diet. The effect of pistachios on insulin and postprandial glucose, endothelial function, and gut hormone has been assessed by Kendall et al. The results demonstrated that the consumption of pistachios reduced the postprandial glycemia, insulin-sparing properties and increased glucagon-like-peptide levels [45].

In one 12-week study of people with metabolic syndrome who ate 1 ounce (25 g) of pistachios twice 2 days, there was a 9% decrease in fasting glucose [51].

5. Conclusion

Nuts are the richest source of healthy nutrients. They are loaded with vitamins, minerals, fibers, and other bioactive compounds. Nuts are nutrients-containing plants. They are used all over the world for nutritional purposes. These have many beneficial effects on health. Bioactive phytochemicals present in nuts are vital for the cognitive function, with positive impact on neurodegeneration and aging. A novel phytochemical present in pistachios, which is called Lutein, provides significant health impacts affecting the cognitive function. Nut consumption reduces the risk of coronary heart diseases, cancer, inflammation, diabetes, and osteoporosis. They are also involved in weight loss.

Author details

Sabira Sultana[1], Muhammad Akram[1*], Nosheen Aslam[2], Naheed Akhter[2], Abdul Wadood Chishti[1], Muhammad Riaz[3], Walaa Fikry Elbossaty[4], Rida Zainab[1] and Umme Laila[1]

1 Department of Eastern Medicine, Faculty of Medical Sciences, Government College University Faisalabad, Pakistan

2 Department of Biochemistry, Government College University Faisalabad, Pakistan

3 Department of Allied Health Sciences, University of Sargodha, Sargodha, Pakistan

4 Department of Biochemistry, Damietta University, Egypt

*Address all correspondence to: makram_0451@hotmail.com

References

[1] Maroney SR. "To Make a Curry the India Way": Tracking the Meaning of Curry Across Eighteenth-Century Communities. UK: Routledge; 2013. pp. 134-146

[2] Wu X, Beecher GR, Holden JM, Haytowitz DB, Gebhardt SE, Prior RL. Lipophilic and hydrophilic antioxidant capacities of common foods in the United States. Journal of Agricultural and Food Chemistry. 2004;**52**(12):4026-4037

[3] Sabate J, Fraser GE, Burke K, Knutsen SF, Bennett H, Lindsted KD. Effects of walnuts on serum lipid levels and blood pressure in normal men. The New England Journal of Medicine. 1993;**328**(9):603-607

[4] Ros E. Health benefits of nut consumption. Nutrients. 2010;**2**(7):652-682

[5] King J, Rechkemmer G, Geiger C. Supplement: 2007 nuts and health symposium. The Journal of Nutrition. 2008;**138**:1734S-1765S

[6] Brufau G, Boatella J, Rafecas M. Nuts: Source of energy and macronutrients. The British Journal of Nutrition. 2006;**96**(S2):S24-SS8

[7] Eaton SB, Konner M. Paleolithic nutrition: A consideration of its nature and current implications. The New England Journal of Medicine. 1985;**312**(5):283-289

[8] Sabaté J. Nut consumption, vegetarian diets, ischemic heart disease risk, and all-cause mortality: Evidence from epidemiologic studies. The American Journal of Clinical Nutrition. 1999;**70**(3):500

[9] Hu FB, Willett WC. Optimal diets for prevention of coronary heart disease. Journal of the American Medical Association. 2002;**288**(20):2569-2578

[10] Nash SD, Nash DT. Nuts as part of a healthy cardiovascular diet. Current Atherosclerosis Reports. 2008;**10**(6):529-535

[11] Lloyd-Jones DM, Hong Y, Labarthe D, Mozaffarian D, Appel LJ, Van Horn L, et al. Defining and setting national goals for cardiovascular health promotion and disease reduction: The American Heart Association's strategic Impact Goal through 2020 and beyond. Circulation. 2010;**121**(4):586-613

[12] Kelly JH, Sabaté J. Nuts and coronary heart disease: An epidemiological perspective. The British Journal of Nutrition. 2006;**96**(S2):S61-SS7

[13] Mukuddem-Petersen J, Oosthuizen W, Jerling JC. A systematic review of the effects of nuts on blood lipid profiles in humans. The Journal of nutrition. 2005;**135**(9):2082-2089

[14] Ros E. Nuts and novel biomarkers of cardiovascular disease. The American journal of clinical nutrition. 2009;**89**(5):1649S-1656S

[15] Mattes RD, Kris-Etherton PM, Foster GD. Impact of peanuts and tree nuts on body weight and healthy weight loss in adults. The Journal of Nutrition. 2008;**138**(9):1741S-1745S

[16] Barreca D, Nabavi SM, Sureda A, Rasekhian M, Raciti R, Silva AS, et al. Almonds (Prunus dulcis Mill. DA webb): A source of nutrients and health-promoting compounds. Nutrients. 2020;**12**(3):672

[17] Casas-Agustench P, Salas-Huetos A, Salas-Salvadó J. Mediterranean nuts:

Origins, ancient medicinal benefits and symbolism. Public Health Nutrition. 2011;**14**(12A):2296-2301

[18] Alasalvar C, Salas-Salvadó J, Ros E, Sabaté J. Health benefits of nuts and dried fruits: An overview. Health Benefits of Nuts and Dried Fruits. 2020:1-10

[19] Bechthold A, Boeing H, Schwedhelm C, Hoffmann G, Knüppel S, Iqbal K, et al. Food groups and risk of coronary heart disease, stroke and heart failure: A systematic review and dose-response meta-analysis of prospective studies. Critical Reviews in Food Science and Nutrition. 2019;**59**(7):1071-1090

[20] Oliveira I, Meyer AS, Afonso S, Aires A, Goufo P, Trindade H, et al. Phenolic and fatty acid profiles, α-tocopherol and sucrose contents, and antioxidant capacities of understudied Portuguese almond cultivars. Journal of Food Biochemistry. 2019;**43**(7):e12887

[21] Barreira JC, Casal S, Ferreira IC, Peres AM, Pereira JA, Oliveira MBP. Supervised chemical pattern recognition in almond (Prunus dulcis) Portuguese PDO cultivars: PCA-and LDA-based triennial study. Journal of Agricultural and Food Chemistry. 2012;**60**(38):9697-9704

[22] Liu Z, Wang W, Huang G, Zhang W, Ni L. In vitro and in vivo evaluation of the prebiotic effect of raw and roasted almonds (*Prunus amygdalus*). Journal of the Science of Food and Agriculture. 2016;**96**(5):1836-1843

[23] Terzo S, Baldassano S, Caldara GF, Ferrantelli V, Lo Dico G, Mulè F, et al. Health benefits of pistachios consumption. Natural Product Research. 2019;**33**(5):715-726

[24] Stone NJ, Robinson JG, Lichtenstein AH, Bairey Merz CN, Blum CB, Eckel RH, et al. ACC/AHA guideline on the treatment of blood cholesterol to reduce atherosclerotic cardiovascular risk in adults: A report of the American College of Cardiology/American Heart Association Task Force on Practice Guidelines. Journal of the American College of Cardiology. 2013, 2014;**63**(25):2889-2934

[25] Poulose SM, Bielinski DF, Shukitt-Hale B. Walnut diet reduces accumulation of polyubiquitinated proteins and inflammation in the brain of aged rats. The Journal of Nutritional Biochemistry. 2013;**24**(5):912-919

[26] Willis LM, Bielinski DF, Fisher DR, Matthan NR, Joseph JA. Walnut extract inhibits LPS-induced activation of BV-2 microglia via internalization of TLR4: Possible involvement of phospholipase D2. Inflammation. 2010;**33**(5):325-333

[27] Nautiyal P. Groundnut. United Nations: Food and Agriculture Organization; 2002

[28] Haddad EH, Gaban-Chong N, Oda K, Sabaté J. Effect of a walnut meal on postprandial oxidative stress and antioxidants in healthy individuals. Nutrition Journal. 2014;**13**:4

[29] Torabian S, Haddad E, Rajaram S, Banta J, Sabaté J. Acute effect of nut consumption on plasma total polyphenols, antioxidant capacity and lipid peroxidation. Journal of Human Nutrition and Dietetics. 2009;**22**(1):64-71

[30] Davis L, Stonehouse W, du Loots T, Mukuddem-Petersen J, van der Westhuizen FH, Hanekom SM, et al. The effects of high walnut and cashew nut diets on the antioxidant status of subjects with metabolic syndrome. European Journal of Nutrition. 2007;**46**(3):155-164

[31] Li Y-C, Qian H, Sun X-L, Cui Y, Wang H-Y, Du C, et al. The effects of germination on chemical composition of peanut seed. Food Science and Technology Research. 2014;**20**(4):883-889

[32] Arya SS, Salve AR, Chauhan S. Peanuts as functional food: A review. Journal of Food Science and Technology. 2016;**53**(1):31-41

[33] Zhao X, Chen J, Du F. Potential use of peanut by-products in food processing: A review. Journal of Food Science and Technology. 2012;**49**(5):521-529

[34] Bonku R, Yu J. Health aspects of peanuts as an outcome of its chemical composition. Food Science and Human Wellness. 2020;**9**(1):21-30

[35] Cordaro M, Siracusa R, Fusco R, D'Amico R, Peritore AF, Gugliandolo E, et al. Cashew (*Anacardium occidentale* L.) Nuts counteract oxidative stress and inflammation in an acute experimental model of carrageenan-induced paw edema. Antioxidants. 2020;**9**:660

[36] Lorenzon Dos Santos J, Quadros AS, Weschenfelder C, Garofallo SB, Marcadenti A. Oxidative stress biomarkers, nut-related antioxidants, and cardiovascular disease. Nutrients. 2020;**12**:682

[37] Ghanavati M, Hosseinabadi SM, Parsa SA, Safi M, Emamat H, Nasrollahzadeh J. Effect of a nut-enriched low-calorie diet on body weight and selected markers of inflammation in overweight and obese stable coronary artery disease patients: A randomized controlled study. European Journal of Clinical Nutrition. 2021;**75**:1099-1108

[38] Aune D, Keum N, Giovannucci E, Fadnes LT, Boffetta P, Greenwood DC, et al. Nut consumption and risk of cardiovascular disease, total cancer, all-cause and cause-specific mortality: A systematic review and dose-response meta-analysis of prospective studies. BMC Medicine. 2016;**14**:207

[39] Sabate J, Ros E, Salas-Salvado J. Nuts: Nutrition and health outcomes. Preface British Journal of Nutrition. 2006;**96**:S1-S2

[40] Parham M, Heidari S, Khorramirad A, Hozoori M, Hosseinzadeh F, Bakhtyari L, et al. Effects of pistachio nut supplementation on blood glucose in patients with type 2 diabetes: A randomized crossover trial. The Review of Diabetic Studies. 2014;**11**:190-196

[41] Mah E, Schulz JA, Kaden VN, Lawless AL, Rotor J, Mantilla LB, et al. Cashew consumption reduces total and LDL cholesterol: A randomized, crossover, controlled-feeding trial. The American Journal of Clinical Nutrition. 2017;**105**:1070-1078

[42] Li H, Li X, Yuan S, Jin Y, Lu J. Nut consumption and risk of metabolic syndrome and overweight/obesity: A meta-analysis of prospective cohort studies and randomized trials. Nutrition and Metabolism. 2018;**15**:46

[43] Hong MY, Groven S, Marx A, Rasmussen C, Beidler J. Anti-Inflammatory, antioxidant, and hypolipidemic effects of mixed nuts in atherogenic diet-fed rats. Molecules. 2018;**23**:3126

[44] Berryman CE, West SG, Fleming JA, Bordi PL, Kris-Etherton PM. Effects of daily almond consumption on cardiometabolic risk and abdominal adiposity in healthy adults with elevated LDL-cholesterol: A randomized controlled trial. Journal of the American Heart Association. 2015;**4**:e000993

[45] Berryman CE, Fleming JA, Kris-Etherton PM. Inclusion of almonds in a cholesterol-lowering diet improves plasma hdl subspecies and cholesterol efflux to serum in normal-weight individuals with elevated LDL cholesterol. The Journal of Nutrition. 2017;**147**:1517-1523

[46] Kendall CW, West SG, Augustin LS, Esfahani A, Vidgen E, Bashyam B, et al. Acute effects of pistachio consumption on glucose and insulin, satiety hormones and endothelial function in the metabolic syndrome. European Journal of Clinical Nutrition. 2014;**68**:370-375

[47] Tan SY, Mattes RD. Appetitive, dietary and health effects of almonds consumed with meals or as snacks: A randomized, controlled trial. European Journal of Clinical Nutrition. 2013;**67**:1205-1214

[48] Dikariyanto V, Berry SE, Francis L, Smith L, Hall WL. Whole almond consumption is associated with better diet quality and cardiovascular disease risk factors in the UK adult population: National Diet and Nutrition Survey (NDNS) 2008-2017. European Journal of Nutrition. 2021;**60**:643-654

[49] Casas R, Sacanella E, Urpí-Sardà M, Chiva-Blanch G, Ros E, Martínez-González MA, et al. The effects of the mediterranean diet on biomarkers of vascular wall inflammation and plaque vulnerability in subjects with high risk for cardiovascular disease. A randomized trial. PLoS One. 2014;**9**:e100084

[50] Mahboobi S. The effect of cashew nut on cardiovascular risk factors and blood pressure: A systematic review and meta-analysis. Current Development in Nutrition. 2019;**3**:nzz031

[51] Parham M, Heidari S, Khorramirad A, Hozoori M, Hosseinzadeh F, Bakhtyari L, et al. Effects of pistachio nut supplementation on blood glucose in patients with type 2 diabetes: A randomized crossover trial. The Review of Diabetic Studies. 2014;**11**(2):190-196

Chapter 2

Quality of Pork Meat Fed with Acorns (*Quercus spp.*)

Paulina G. Flores

Abstract

Pork is one of the most popular foods in Chile. Consumers are concerned about their nutritional quality and about the characteristics of the production systems. Outdoor production systems are the most valued by consumers due to animal welfare conditions related to the nutritional quality of the product. On the other hand, intensive or traditional confined pig production systems represent an environment that alters the nutrients in meat products. For this reason, outdoor production systems become more important, especially when nuts (Quercus spp.) are included in pig feeding. Acorn nuts have a high level of fatty acids transferred to the meat, increasing flavor and juiciness. This proposal is a bibliographic review that will also include the advantage of controlling an invasive species, such as wild boar for a small producer in southern Chile, who could obtain a meat product with healthy fatty acids, favoring the sustainability of the ecosystem.

Keywords: feed, nuts, meat quality, pig production system, wild boar

1. Introduction

Pork meat is one of the most popular food products in Chile. Its consumption reaches a value close to 23% of the total meat consumed. Consumers of meat products are increasingly interested in the nutritional quality of meat, and pork has particular attention given the implications for people's health [1]. Feed quality influences the fat intramuscular fat content of the pig, with some aspects of quality decreasing as the composition of fatty acid in intramuscular lean carcass content [2]. Compared to extensive and natural systems, the inferior quality of pork from intensive systems presents quality defects such as exudates from meat juices and a decrease in the amount of intramuscular fat [3].

2. Meat composition

The composition of meat comprises moisture, protein, fat, minerals, and a small proportion of carbohydrates. Chemically, a piece of lean meat contains approximately

Meat products[a]	Energy (kcal)	Prot[b] (g)	Fat (g)	Sat[c] fat (g)	Cl[d] (mg)	Vit[e] B12 (mcg)	Fe[f] (mg)	Zn[g] (mg)
Meat, raw/unprepared								
Beef ribeye (*L*[h]*. thoracis*)	166	17.51	22.07	9000	68	3.11	1.87	3.85
Beef strip loin (*L. lumborum*)	228	20.61	15.49	6251	81	1.06	1.47	3.53
Beef tenderloin (*Psoas major*)	247	19.61	18.16	8410	85	1.01	1.42	2.90
Beef top sirloin (*Pluteus medius* et al.)	201	20.30	12.71	5127	75	1.05	1.48	3.55
Beef tri-tip (*m. tensor fasciae,latae*)	142	21.26	5.63	1729	61	1.04	1.54	3.85
Beef eye round (*Semitendinosus*)	124	23.27	3.44	1276	60	1.83	1.45	3.40
Beef brisket (*Deep pectoral et al.*)	132	21.47	5.11	1844	67	1.81	2.06	5.21
Beef flank (*Rectus abdominis*)	155	21.22	7.17	2978	65	1.09	1.55	3.70
Beef, ground 80% lean, 20% fat	254	17.17	20.00	7581	71	2.14	1.94	4.18
Beef, ground 90% lean, 10% fat	176	20.00	10.00	3927	65	2.21	2.24	4.79
Beef, ground 93% lean, 7% fat	152	20.85	7.00	2878	63	2.23	2.33	4.97
Beef, ground 97% lean, 3% fat	121	21.98	3.00	1480	60	2.26	2.44	5.21
Pork loin (*L. lumborum*)	198	19.74	12.58	4360	63	0.53	0.79	1.74
Pork ham (*Biceps femoris et al.*)	245	17.43	18.87	6540	73	0.63	0.85	1.93
Pork, ground 84% lean, 16% fat	218	17.99	16.00	4930	68	0.73	0.88	1.91
Pork, ground 96% lean, 4% fat	121	21.10	4.00	1420	59	0.64	0.86	1.93

[a]*All products are standardized to a 100 g serving*
[b]*Prot: Protein*
[c]*Sat: Saturated*
[d]*Cl: Cholesterol*
[e]*Vit: Vitamin*
[f]*Fe: Iron*
[g]*Zn: Zinc*
[h]*L: Longissimus*

Table 1.
Nutritional composition of different food products [6].

72% water, 21% protein, 5% fat, and 1% ash. The macronutrient protein is the most valuable component in nutrition and processing [4]. In addition, it is a source of essential amino acids, vitamin B12, iron, and zinc [5]. It is low in carbohydrates and does not contain dietary fiber (**Table 1**).

2.1 Acorn characteristics

Quercus robur L., the common oak, is a plant in the family Fagaceae [7] composed of 300 species around the world, including deciduous and evergreen trees [8]. The Quercus genus can be located in America, temperate Europe, Asia, and subtropical Africa [9]. This tree, whose fruit is known as "bellota" has different shapes and varies in size [10]. Because of their content and nutritional importance, they have been called "secondary human food," being part of the human diet for centuries, as raw material in bread making [8, 11].

In Europe, the Quercus tree is considered the main tree species in floodplain forests of central Europe [8]. In the Czech Republic, the clear-cut silvicultural system represents a predominant method of regeneration of oak stands in a flooded forest [11].

The consumption of acorns in human food has a long history. The population of Italy and Spain consumes this fruit.

Using acorns in human nutrition has a long history. In Italy and Spain, it has been part of the local diet for some time, contributing up to 25% of consumed food by the poorest classes as a substitute for coffee [12]. Algeria, Morocco, and the US used acorn oil as a cooking oil and medicine for burns and injuries [13].

Acorn oils have similar properties to olive oil, such as color, iodine value, UV extinction coefficient, fatty acid composition, and refractive index [14, 15]. In South America, the *Q. robur* tree is part of the forests of Southern Chile, including Chilean Patagonia and Argentina. The fruits of this tree are part of the diet of wild boars and domestic pigs, and it is used as an ornamental tree species in squares in different cities (**Figure 1**).

Acorn's nutritional composition is rich in unsaturated fatty acids (75–90%), especially oleic acid (65%) and linoleic acid. These unsaturated fatty acids are essential for synthesizing eicosanoids, favoring the reduction of blood triglycerides and the increase of HDL cholesterol [16]. Compared to commercial foods, acorns have a higher content of crude fat, net energy, and metabolizable energy [17] (**Table 2**).

In Europe, acorns are used as a feed source for free-range animals, especially Iberian pigs, during the montanera season [18]. Montanera is the fattening period of Iberian pigs that weigh more than 40 kg in two or three months. In the Iberian Peninsula, the Iberian pig is an autochthonous breed. The Iberian pig system is linked to the dehesa lands of *Quercus* spp., in the Iberian Peninsula [19]. The abundant food provided by acorns is used to fatten the pigs from early November to late February [20].

In the dehesa, the pig diet is based upon acorns and grass, the most abundant resources in the area during this period [21]. Grazing is a repeated ingestion of small quantities of food [20]. Grass bites can be very light, but pigs need a lot of them as a source of protein to compensate for the low concentration of this nutrient in the acorn.

Figure 1.
Oak tree (Quercus robur) in the square of Coyhaique city, Chilean Patagonia.

Pigs have a good sense of smell and taste [22]. They select their food depending on crude protein content and essential amino acids such as lysine, methionine, and tryptophan [23]. This sensory ability allows pigs to differentiate between acorn fruits from different oak trees [24].

The feeding systems for pigs, whose feeding base is the acorn fruit, contribute to delivering characteristics to the carcass and the meat, especially in the composition of fatty acids of the adipose tissues [20].

2.2 Fatty acids composition and other compounds in muscle as an effect of acorns feeding

The fatty acid composition of the diet reflects the quality of the tissue. Animals fed n-3 polyunsaturated fatty acids (PUFA) are a source of fatty acids in the human diet [25]. That is, by consuming meat from pigs fed on acorns and plants rich in n-6:n-3 fatty acids [26]. Several meat products reflect the diet of animals, such as fresh meats, sausages, cured meats, hams, and smoked meats, providing an unlimited range of highly appreciated flavors and aromas [27].

The evidence of an exclusive feeding of acorns in the animal diet is consistent in showing a higher lipid content (81.69% of total fatty acids) while feeding pigs with

Nutrient composition	Feedstuff	Fresh acorns	P^*
Moisture (g/100 g)	14^b	32.4^a	0.0017
Crude protein (g/100 g)	15.6^a	5.6^b	0.0001
Crude fat (g/100 g)	3.3^b	5^a	0.002
Fibrer (g/100 g)	6.8^a	5.6^b	0.013
Ash (g/100 g)	3.7^a	2.7^b	0.010
Net energy (Mcal/kg)	1.73^b	1.85^a	0.09
Metabolizable energy (Mcal/kg)	2.9^b	3.13^a	0.011
Value D	80^b	88^a	0.0012
		Fatty acid profile mg/100 mg	
C 14:0	18.16^a	0.12^b	0.0032
C16:0		17.48	ns^A
C 16:1*n-9*	1.18	1.7	ns
C 16:1*n-7*		0.16	
C 17:0		0.05	
C18:0		1.72	
C18:1*n-9*	18.39^b	27.3^a	0.0016
C18:1*n-7*	0.75^b	5.84^a	0.0031
C18:2*n-6*	57.7^a	39.1^b	0.009
C 20:0	3.51^b	4.58^a	0.0001
C18:3*n-3*		1.65	
C 20:2	19.75^a	0.06b	0.0001
SFA^B		23.9	
$MUFA^C$	19.1^b	35.01^a	0.0001
$PUFA^D$	61.2^a	40.8^b	0.0001
$n6{:}n3^E$	16.4^b	23.7^a	0.001
$P{:}S^F$	3.1a	1.7b	0.001

[A]ns: not significant
[B]Σ Saturated fatty acids: C14 + C16 + C17 + C18 + C20
[C]Σ Monounsaturated fatty acids: C16:1 + C18:1n-9 + C18:1n-7
[D]Σ Polyunsaturated fatty acids: C18:2n-6 + C18:3n-3 + C20:2
[E]n-6:n-3: Fatty acids ratio (linoleic acid/linolenic acid)
[F]P:S: Polyunsaturated: Saturated fatty acids ratio
**The p-value corresponds to the Student t-test between feedstuff and fresh acorn*

Table 2.
Analyzed composition of feed (g/100 g) and fatty acid (mg/100 mg) [17] composition.

grass and chestnuts (78.80% and 77. 62% of total fatty acids, respectively) have a lower fatty acid value [28, 29].

The highest percentages of monounsaturated fatty acids (MUFA), especially C18:1n-9, have also been found in diets rich in acorns [29], as well as in ham and loin muscles of Iberian pigs fed exclusively with acorns and grass in the fattening of montanera systems [30, 31].

Wild boars (*Sus scrofa*) raised with acorns had significantly ($P = 0.0001$) higher proportions of C18:1 *n-9* in *Longissimus lumborum* (LL) muscle than wild boars feeding with commercial concentrate, based mainly on corn [17]. There is a significant difference between pigs under free-range systems and confinement, as showed in **Table 3**.

According to Ref. [33], the hypothesis is that fatty acids composition depends on muscle type (oxidative or glycolytic), where a fasted period can be sufficient to decrease MUFA and n-6, but Ref. [26] suggests that a new experiment must have been made to verify this hypothesis.

The antioxidant composition effect also has been widely documented as the influence of feeding on the tocopherol content in muscle. γ tocopherol vitamin content of muscle acquired from free-range pigs is recognized in acorn feeding diets [31, 34]. The meat from these pigs is evidence of high quality [21]. Diet used in finishing phase (up to 155 kg body weight) is based on grass and acorn without restrictions [30, 31].

Chemical and antioxidant composition increase energy and fat levels in feed [17]. Some researchers showed negative protein digestibility because of acorn shells, especially on lysine amino acids [35–37].

It is important to consider that the values realized in the previous table, although they show high values of fatty acids, especially those MUFA and PUFA that contribute

Fatty acid composition	Wild boar ()	Iberian pig ()
C 14:0	1.22	1.31
C16:0	23.51	24.19
C 16:1*n-9*	0.42	—
C 16:1*n-7*	2.69	—
C 17:0	0.33	0.14
C18:0	12.1	10.22
C18:1*n-9*	41.1	51.39
C18:1*n-7*	3.11	—
C18:2*n-6*	13.1	5.41
C 20:0	0.19	0.17
C18:3*n-3*	0.51	0.37
C 20:2	0.57	0.21
SFA[a]	38.77	36.02
MUFA[b]	48.11	57.13
PUFA[c]	14.47	6.85
n6:n3[d]	20.76	—
P:S[e]	0.39	—

[a]*Σ Saturated fatty acids: C14 + C16 + C17 + C18 + C20.*
[b]*Σ Monounsaturated fatty acids: C16:1 + C18:1n-9 + C18:1n-7.*
[c]*Σ Polyunsaturated fatty acids: C18:2n-6 + C18:3n-3 + C20:2.*
[d]*n-6:n-3: Fatty acids ratio (linoleic acid/linolenic acid).*

Table 3.
Fatty acid composition (mg/100 mg) of L. lumborum muscle from wild boars [17] and Iberian pigs [32] fed with acorns.

	Control	20%	40%	R^2
Flavor	4.99[c]	5.04[b]	5.33[a]	0.620
Tenderness	4.43[c]	5.05[b]	5.22[a]	0.812
Juiciness	4.4[c]	4.71[b]	5[a]	0.849

[*]Values with different superscripts are significantly different using the Duncan test (P < 0.05).

Table 4.
Sensory characteristics of L. lumborum muscle from wild boars fed the experimental diets.

Karyotype	Cholesterol [41]	Cholesterol [42]
2n = 36	29.60 ± 6.6	20.9 ± 4.52
2n = 37	62.58 ± 17.5	36.8 ± 5.5
2n = 38	28.94b ± 5.3	37.2 ± 8.2

Table 5.
Cholesterol content in Semimembranosus muscle of the different karyotype groups (mg/100 g).

to the concentration of essential fatty acids, come from intensive systems, which are becoming more and more relegated from the point of view of the consumer. In addition, it should be noticed that the free-range systems, where the Iberian pig predominates, also get outstanding values in meat at levels of essential fatty acids. The high sensory value from pig meat under extensive systems is attributed to essential fatty acids [38], especially in dry cured products with a strong flavor [19, 39, 40].

Table 4 shows a descriptive sensory analysis made using a hedonic scale to reflect the perception by consumers of meat sensory characteristics from wild board fed with 20% and 40% of acorn diet inclusion. The Control group received a commercial concentrate base in corn. The consumers detected a positive relationship between the percentages of MUFA and PUFA with flavor, tenderness, and juiciness attributes. This is explained by the evidence that meat from animals fed with acorns was perceived as juicier and more tender by a sensory panel [17].

Differences in meat from wild boar compared to pork have been reported in particular considering cholesterol contents. In Brazil, Ref. [41] analyzed cholesterol in the meat of wild boar (2n = 36) and crossbreeds (2n = 37 y 38) founding lower cholesterol values in wild boards. Similar results were found in Ref. **Table 5** [42].

3. Other values from feeding with acorn and sustainability systems

Meat products derived from animals raised under extensive, free-range conditions, mainly feeding with acorns and grass in silvopastoral systems, as occurs, for example, in the Spanish system called montanera, comply with these requirements [32], and their high content of essential fatty acids makes it healthier than other animal products [43]. The extensive, free-range conditions contribute to the conservation of the dehesa ecosystem, where the extensive grazing system is a major determinant of biodiversity and ecosystem functioning in a forested Mediterranean grassland environment [44]. The environmental and economic importance of dehesas is due to their use in extensive free range and also to their characteristic high biodiversity [45]. This makes sense when the use of a site is done in a controlled manner, especially when it is already known that grazing can have different effects on soil fertility and the communities of microorganisms, plants, and animals that develop from these soils depend on the local climate, topography, and physicochemical properties of the parent soil material [46], affecting soil fertility thought trampling and dung generating an easily litter degradation [47, 48].

The animals also remove the plant biomass, encouraging fast-growing species with high content of nutrients on their leaves and palatability [49]. On the opposite, the abandonment of free-range grazing leads to the existence of marginal areas grazing affects grassland dynamics, changing the species colonization and extinction relationships via consumption, mechanical disturbance, seed dispersal, and altered soil fertility due to dung-borne nutrient input [50]. Also, these sites are associated with a reduction in soil fertility-associated variables such as organic matter, total nitrogen, the availability of phosphorus [51], soil disturbance, carbon emission, and soil degradation [52].

There is an important point that stands out in the publication of Ref. [53], which is related to the sustainability of ecosystems, such as human impacts on the soil that remain unexplored. One potential source of anthropogenic soil carbon emissions is invasive species, with research suggesting that human-propagated animals ranging from insects such as mountain pine beetles and earthworms to herbivorous mammals, such as cattle, deer, goats, rabbits, and pigs, cause serious disturbances to soil properties through loss of vegetation and direct soil disturbance. They also evaluated the global ramifications of one of the most widely spread human species, wild pigs (*Sus scrofa*), on soil disturbance and CO2 emissions.

This makes sense since, although in Chile, wild boars are raised for commercial intentions in confinement conditions [54], however, an important part of the population lives free on the edge of the Andes Mountain range, from the region of Araucania to the Aysen region, in the heart of Chilean Patagonia [55].

The wild boar, an invader of South America [56], arrives in Chile in two ways, a direct importation from Germany and subsequent release in 1952 in the Villarrica National Park, and then by specimens that by their own means pass from Argentina from 1958 onwards [57]. It is an exotic species little studied, and only recently has it been published about its diet in southern America, focusing attention on the predation of birds and seeds of endemic trees (**Figure 2**) [55].

Their presence has a powerful impact, and often they could be responsible for the impoverishment of the local flora and fauna, as has happened in other places with invasive animal species [58, 59]. Management of wild boar population size in Chile is predominantly carried out by hunting at bait sites or by hunting dogs. The capture of live traps to latest keep them in confinement for meat production is another way of controlling the population. Wild boars and domestic pigs becoming from *Sus scrofa domesticus*. Wild boar can be reared in captivity.

The information found on the benefits of acorn-fed wild boar meat and the antecedents on the breeding of the Iberian pig in montaneras and Dehesas, very widespread agrosilvopastoral ecosystems in the Mediterranean ecosystems of the Iberian Peninsula, reveals an important alternative for the control of a species invasive within the Patagonian ecosystems. The economic and environmental importance of the Dehesas is because of the use of extensive livestock farming and also to their characteristic high biodiversity, the same livestock environment declared for the silvopastoral system's characteristic of southern Chile, where the wild boar lives.

Wild boar causes serious impacts on biodiversity and ecosystems in silvopastoral systems in Patagonia by affecting physicochemical soil and water bodies properties, plant diversity, predating on wild animals, seeds, and seedlings, destroying bird

Figure 2.
Wild boar Sus scrofa scrofa from Chillan city, Chile.

nests, and interacting with fauna by competition, hybridization, and transmission of *Toxoplasma gondii* and *Trichinella spp.*, zoonotic and commercially important diseases [60–62] found that wild boar can cause economic losses in production activities because of the consumption of livestock offspring sprouts that feed cattle. In this way, the interspecific interaction between wild boars and other species can affect native ecosystems, but more studies are needed [63].

The incorporation of *Q. robur* into the Argentine Patagonian forest has been reported, standing out for its high productivity in forest environments where attempts have been made to diversify production systems and thus maintain the sustainability of the ecosystem. According to [64], *Q. robur*, at 150 years of age, reaches 25–30 m in height in the best sites in central Europe. In Patagonia, at 45–50 years, it reaches dominant heights of 20 to 24 m, which indicates that this species' turn would be well advanced.

There is a known conflict in the management of invasive species to conserve the ecological, social, and cultural values of the invaded ecosystems in complement to the use or utility of exotic species in productive activities [65]. As suggested by [63], management has become a complex issue that requires an interdisciplinary study with multiple approaches to develop an adaptive management strategy.

The vision of control of this species for economic and sustainable purposes of the environment will be an alternative to a meat organic farm system in an environment where extensive livestock is the pillar of the Patagonian agricultural tradition, and wild boars are one of the most damaging invasive species in the world, with a significant impact on the Patagonian forest degradation [66].

Meat production from wild boars in countries such as Chile has been increasing over the last decade, with an expanding demand both in Chile and overseas [67], for its benefits of being a meat product with healthy fatty acids and an alternative to contribute to the sustainability of the ecosystem.

4. Conclusions

There is great interest in the consumption of animal meat from free-range systems. Pork is very popular and generally provides an important source of nutrients, mainly protein. When pigs receive acorns, the fruit of the oak *Q. robur*, in their diet, they showed an increase in the level of essential fatty acids. Examples of this are the Iberian pig in the Spanish Iberian province and the wild boar in southern Chile. Both show an increase in the level of fatty acids *n3* and *n6* and a decrease in cholesterol when acorns are part of their diet.

The Iberian pig is endemic in Spain, the wild boar is an introduced species that harm the ecosystem. Therefore, its use as a productive species could contribute as a measure of control of the species and also contribute to the sustainability of the environment.

Acknowledgements

The author thanks the Government of the Chile Ministry of Education, URY project number 21991 Mineduc, Innovation, Technology, and Transfer Unit, Research Direction, for the financial support of the present book chapter. The author also wishes to acknowledge the University of Aysen for its support during the writing of this manuscript. The content of this chapter reflects only the author's view. The Government of Chile, the Ministry of Education, the University of Aysen, and the Innovation, Technology, and Transfer Unit are not responsible for any use that may be made of the information it contains.

Conflict of interest

The authors declare no conflict of interest.

Author details

Paulina G. Flores
University of Aysen, Coyhaique, Chile

*Address all correspondence to: paulina.flores@uaysen.cl

References

[1] Raes K, De Smet S, Demeyer D. Effect of dietary fatty acids on incorporation of long chain polyunsaturated fatty acids and conjugated linoleic acid in lamb, beef and pork meat: A review. Animal Feed Science and Technology. 2004;**113**:199-221. DOI: 10.1016/j.anifeedsci.2003.09.001

[2] Cameron ND, Enser M, Nute GR, Whittington FM, Penman JC, Fisken AC, et al. Genotype with nutrition interaction on fatty acid composition of intramuscular fat and the relationship with flavour of pig meat. Meat Science. 2000;**55**:187-195. DOI: 10.1016/S0309-1740(99)00142-4

[3] Wojtysiak D, Połtowicz K. Carcass quality, physico-chemical parameters, muscle fibre traits and myosin heavy chain composition of m. longissimus lumborum from Puławska and Polish Large White pigs. Meat Science. 2014;**97**:395-403

[4] Pereira P, Vicente A. Meat nutritional composition and nutritive role in the human diet. Meat Science. 2013;**93**:586-592. DOI: 10.1016/j.meatsci.2012.09.018

[5] McAfee AJ, McSorley EM, Cuskelly GJ, Moss BW, Wallace JM, Bonham MP, et al. Red meat consumption: An overview of the risks and benefits. Meat Science. 2010;**84**:1-13. DOI: 10.1016/j.meatsci.2009.08.029

[6] USDA, AMS. 2016. U.S. Department of Agriculture, Agricultural Marketing Service Retail reports (2016) Retrieved 2016 December from: https://www.ams.usda.gov/market-news/retail

[7] Antov MG, Šćiban MB, Prodanović JM, Kukić DV, Vasić VM, Đorđević TR, et al. Common oak (Quercus robur) acorn as a source of natural coagulants for water turbidity removal. Industrial Crops and Products. 2018;**117**:340-346. DOI: 10.1016/j.indcrop.2018.03.022

[8] Brischke C, Welzbacher CR, Rapp AO, Augusta U, Brandt K. Comparative studies on the in-ground and above-ground durability of European oak heartwood (Quercus petraea Liebl. and Quercus robur L.). European Journal of Wood and Wood Products. 2009;**67**:329-338. DOI: 10.1007/s00107-009-0329-x

[9] Shahbaz SE, Abdulrahman SS, Abdulrahman HA. Use of leaf anatomy for identification of Quercus L. species native to Kurdistan-Iraq. Science Journal of University of Zakho. 2015;**3**:222-232

[10] Pritchard HW, Daws MI, Fletcher BJ, Gaméné CS, Msanga HP, Omondi W. Ecological correlates of seed desiccation tolerance in tropical African dryland trees. American Journal of Botany. 2004;**91**:863-870. DOI: 10.3732/ajb.91.6.863

[11] Martiník A, Dobrovolny L, Palátová E. Tree growing space and acorn production of Quercus robur. Dendrobiology. 2014:71. DOI: 10.12657/denbio.071.010

[12] Hill AF. Economic Botany: A Textbook of Useful Plants and Plant Products. 2nd ed. New York: McGraw-Hill; 1952. p. 560

[13] Taib M, Bouyazza L, Lyoussi B. Acorn oil: Chemistry and functionality. Journal of Food Quality. 2020;**2020**:1-11. DOI: 10.1155/2020/8898370

[14] Deforce K, Bastiaens J, Van Calster H, Vanhoutte S. Iron age acorns

from Boezinge (Belgium): The role of acorn consumption in prehistory. Archaologisches Korrespondenzblatt. 2009;**12**:381-392

[15] Vinha AF, Barreira JC, Costa AS, Oliveira MBPA. New age for Quercus spp. fruits: Review on nutritional and phytochemical composition and related biological activities of acorns. Comprehensive Reviews in Food Science and Food Safety. 2016;**15**:947-981

[16] Uddin G, Rauf A. Phytochemical screening, antimicrobial and antioxidant activities of aerial parts of *Quercus robur* L. Middle-East Journal of Medicinal Plants Research. 2012;**1**:1-4

[17] Flores PG, Morales R, Skewes O. Chemical properties and sensory characteristics of wild boar meat (Sus scrofa scrofa) fed with acorns (*Quercus robur*). Revista de Producción. Animal. 2020;**33**:1-12

[18] Cantos E, Espín JC, López-Bote C, de la Hoz L, Ordóñez JA, Tomás-Barberán FA. Phenolic compounds and fatty acids from acorns (Quercus spp.), the main dietary constituent of free-ranged Iberian pigs. Journal of Agricultural and Food Chemistry. 2003;**51**:6248-6255

[19] Ayuso D, González A, Peña F, Hernández-García FI, Izquierdo M. Effect of fattening period length on intramuscular and subcutaneous fatty acid profiles in Iberian pigs finished in the montanera sustainable system. Sustainability. 2020;**12**:7937. DOI: 10.3390/su12197937

[20] Rodríguez-Estévez V, García A, Peña F, Gómez AG. Foraging of Iberian fattening pigs grazing natural pasture in the dehesa. Livestock Science. 2009;**120**:135-143. DOI: 10.1016/j.livsci.2008.05.006

[21] Szyndler-Nędza M, Świątkiewicz M, Migdał Ł, Migdał W. The quality and health-promoting value of meat from pigs of the native breed as the effect of extensive feeding with acorns. Animals. 2021;**11**:789. DOI: 10.3390/ani11030789

[22] Roura E, Humphrey B, Tedó G, Ipharraguerre I. Unfolding the codes of short-term feed appetence in farm and companion animals. A comparative oronasal nutrient sensing biology review. Canadian Journal of Animal Science. 2008;**88**:535-558

[23] Kyriazakis I, Emmans GC, Whittemore CT. Diet selection in pigs: Choices made by growing pigs given foods of different protein concentrations. Animal Science. 1990;**51**:189-199. DOI: 10.1017/S1357729800050712

[24] Canelo T, Pérez-Izquierdo C, Gaytán Á, Bonal R. Intraguild predation of weevils by livestock reduces acorn pests in oak silvopastoral systems. Journal of Pest Science. 2021;**94**:541-551. DOI: 10.1007/s10340-020-01278-8

[25] Pal J, Shukla BN, Maurya AK, Verma HO, Pandey G, Amitha A. A review on role of fish in human nutrition with special emphasis to essential fatty acid. International Journal of Fisheries and Aquatic Studies. 2018;**6**:427-430

[26] de Tonnac A, Guillevic M, Mourot J. Fatty acid composition of several muscles and adipose tissues of pigs fed n-3 PUFA rich diets. Meat Science. 2018;**140**:1-8. DOI: 10.1016/j.meatsci.2017.11.023

[27] Michel F, Hartmann C, Siegrist M. Consumers' associations, perceptions and acceptance of meat and plant-based meat alternatives. Food Quality and Preference. 2021;**87**:104063. DOI: 10.1016/j.foodqual.2020.104063

[28] Rey AI, Daza A, López-Carrasco C, López-Bote CJ. Quantitative study of

the tocopherols accumulation in muscle and backfat from Iberian pigs kept free-range as affected by time of free-range feeding or weight gain. Animal Science. 2006;**82**:901-908. DOI: 10.1017/ASC2006113

[29] Pugliese C, Sirtori F, Ruiz J, Martin D, Parenti S, Franci O. Effect of pasture on chestnut or acorn on fatty acid composition and aromatic profile of fat of Cinta Senese dry-cured ham. Grasas y aceites. 2009;**60**:271-276. DOI: 10.3989/gya.130208

[30] Lebret B, Lenoir H, Fonseca A, Riquet J, Mercat M. Finishing season and feeding resources influence the quality of products from extensive-system gascon pigs. Part 2: Muscle traits and sensory quality of dry-cured ham. Animal. 2021;**15**:100305. DOI: 10.1016/j.animal.2021.100305

[31] Tejerina D, García-Torres S, De Vaca MC, Vázquez FM, Cava R. Effect of production system on physical–chemical, antioxidant and fatty acids composition of longissimus dorsi and Serratus ventralis muscles from Iberian pig. Food Chemistry. 2012;**133**:293-299. DOI: 10.1016/j.foodchem.2012.01.025

[32] Tejeda JF, Hernández-Matamoros A, Paniagua M, González E. Effect of free-range and low-protein concentrated diets on growth performance, carcass traits, and meat composition of Iberian pig. Animals. 2020;**10**:273. DOI: 10.3390/ani10020273

[33] Malmfors B. Fatty acid composition of porcine back fat and muscle lipids as affected by sex, weight and anatomical location. Sweden Journal of Agriculture Research. 1978;**8**:25-38

[34] Ortiz A, Tejerina D, Díaz-Caro C, Elghannam A, García-Torres S, Mesías FJ, et al. Is packaging affecting consumers' preferences for meat products? A study of modified atmosphere packaging and vacuum packaging in Iberian dry-cured ham. Journal of Sensory Studies. 2020;**35**:e12575. DOI: 10.1111/joss.12575

[35] Marín-García PJ, Llobat L. How does protein nutrition affect the epigenetic changes in pig? A review. Animals. 2021;**11**:544. DOI: 10.3390/ani11020544

[36] Liao SF, Wang T, Regmi N. Lysine nutrition in swine and the related monogastric animals: Muscle protein biosynthesis and beyond. Springerplus. 2015;**4**:1-12. DOI: 10.1186/s40064-015-0927-5

[37] Eugenio FA, Van Milgen J, Duperray J, Sergheraert R, Floc'h L. Feeding intact proteins, peptides, or free amino acids to monogastric farm animals. Amino Acids. 2022:1-12

[38] Mekki I, Smeti S, Hajji H, Yagoubi Y, Mahouachi M, Atti N. Effect of oak acorn (Quercus ilex) intake during suckling and fattening of Barbarine lambs on growth, meat quality and fatty acid profile. Journal of Animal Feed Science. 2019;**28**:22-30. DOI: 10.22358/jafs/102757/2019

[39] Perez-Santaescolastica C, Fraeye I, Barba FJ, Gomez B, Tomasevic I, Romero A, et al. Application of non-invasive technologies in dry-cured ham: An overview. Trends in Food Science & Technology. 2019;**86**:360-374. DOI: 10.1016/j.tifs.2019.02.011

[40] Carrapiso AI, Tejeda JF, Noguera JL, Ibáñez-Escriche N, González E. Effect of the genetic line and oleic acid-enriched mixed diets on the subcutaneous fatty acid composition and sensory characteristics of dry-cured shoulders from Iberian pig. Meat Science. 2020;**159**:107933. DOI: 10.1016/j.meatsci.2019.107933

[41] Lui JF, Macuco VSO, Neto AC, Tosta PA, Malheiros EB. Lipídeo, proteína e colesterol na carne de javalis (Sus scrofa scrofa) de diferentes grupos genéticos. Archivos de zootecnia. 2007;**56**:951-954

[42] Skewes O, Morales R, Mendoza N, Smulders FJM, Paulsen P. Carcass and meat quality traits of wild boar (Sus scrofa s. L.) with 2n= 36 karyotype compared to those of phenotypically similar crossbreeds (2n= 37 and 2n= 38) raised under the same farming conditions 2: Fatty acid profile and cholesterol. Meat Science. 2009;**83**:195-200. DOI: 10.1016/j.meatsci.2009.04.017

[43] González-Domínguez R, Sayago A, Fernández-Recamales Á. Fatty acid profiling for the authentication of Iberian hams according to the feeding regime. Food. 2020;**9**:149. DOI: 10.3390/foods9020149

[44] Oggioni SD, Ochoa-Hueso R, Peco B. Livestock grazing abandonment reduces soil microbial activity and carbon storage in a Mediterranean Dehesa. Applied Soil Ecology. 2020;**153**:103588. DOI: 10.1016/j.apsoil.2020.103588

[45] Hoffmann I, Boerma D. Ecosystem Services Provided by Livestock Species and Breeds, with Special Consideration to the Contributions of Small-scale Livestock Keepers and Pastoralists. Rome: FAO; 2014

[46] Eldridge DJ, Delgado-Baquerizo M. Grazing reduces the capacity of landscape function analysis to predict regional-scale nutrient availability or decomposition, but not total nutrient pools. Ecological Indicators. 2018;**90**:494-501. DOI: 10.1016/j.ecolind.2018.03.034

[47] Pietrosemoli S, Raczkowski C, Green JT Jr, Villamide MJ. Effects of growing-finishing pig stocking rates on bermudagrass ground cover and soil properties. Animals. 2020;**10**(9):1666. DOI: 10.3390/ani10091666

[48] Sun B, Overbury T, Hopkins DW. Soil carbon and nutrient variations in an arable-ley rotation with organic pig production. Biological Agriculture & Horticulture. 2022;**38**(1):61-71. DOI: 10.1080/01448765.2021.1991845

[49] De Deyn GB, Cornelissen JH, Bardgett RD. Plant functional traits and soil carbon sequestration in contrasting biomes. Ecology Letters. 2008;**11**:516-531. DOI: 10.1111/j.1461-0248.2008.01164.x

[50] Peco B, Sánchez AM, Azcárate FM. Abandonment in grazing systems: Consequences for vegetation and soil. Agriculture, Ecosystems & Environment. 2006;**113**:284-294. DOI: 10.1016/j.agee.2005.09.017

[51] Vinograd A, Zaady E, Kigel J. Dynamics of soil nutrients in abandoned sheep corrals in semi-arid Mediterranean planted forests under grazing. Journal of Arid Environments. 2019;**164**:38-45. DOI: 10.1016/j.jaridenv.2019.02.007

[52] Sanderman J, Hengl T, Fiske GJ. Soil carbon debt of 12,000 years of human land use. Proceedings of the National Academy of Sciences. 2017;**114**:9575-9580. DOI: 10.1073/pnas.1706103114

[53] O'Bryan CJ, Patton NR, Hone J, Lewis JS, Berdejo-Espinola V, Risch DR, et al. Unrecognized threat to global soil carbon by a widespread invasive species. Global Change Biology. 2022;**28**:877-882

[54] Skewes O, Morales R. Wild boar (*Sus scrofa* L.) breeding in Chile. Basic aspects of distribution, size and management. Agro-Ciencia. 2006;**22**:29-36

[55] Skewes O, Moraga CA, Arriagada P, Rau JR. El jabalí europeo (Sus scrofa): Un invasor biológico como presa reciente del puma (Puma concolor) en el sur de Chile. Revista chilena de historia natural. 2012;**85**:227-232. DOI: 10.4067/S0716-078X2012000200009

[56] Jaksic FM, Iriarte JA, Jiménez JE, Martínez DR. Invaders without frontiers: Cross-border invasions of exotic mammals. Biological Invasions. 2002;**4**:157-173. DOI: 10.1023/A:1020576709964

[57] Skewes O. Status des Wildschweines, Sus scrofa L., in Chile [thesis]. Dokumentation einer Fremdansiedlung von europäischem Wild; 1990

[58] Courchamp F, Chapuis JL, Pascal M. Mammal invaders on islands: Impact, control and control impact. Biological Reviews. 2003;**78**:347-383. DOI: 10.1017/S1464793102006061

[59] Fahlman Å, Lindsjö J, Norling TA, Kjellander P, Ågren EO, Bergvall UA. Wild boar behaviour during live-trap capture in a corral-style trap: Implications for animal welfare. Acta Veterinaria Scandinavica. 2020;**62**:1-11. DOI: 10.1186/s13028-020-00557-9

[60] Barrios-Garcia MN, Ballari SA. Impact of wild boar (Sus scrofa) in its introduced and native range: A review. Biological Invasions. 2012;**14**:2283-2300. DOI: 10.1007/s10530-012-0229-6

[61] Winter M, Abate SD, Pasqualetti MI, Fariña FA, Ercole ME, Pardini L. Toxoplasma gondii and Trichinella infections in wild boars (Sus scrofa) from Northeastern Patagonia, Argentina. Preventive Veterinary Medicine. 2019;**168**:75-80. DOI: 10.1016/j.prevetmed.2019.04.014

[62] Bueno CG, Barrio IC, García-González R, Alados CL, Gómez-García D. Does wild boar rooting affect livestock grazing areas in alpine grasslands? European Journal of Wildlife Research. 2010;**56**:765-770. DOI: 10.1007/s10344-010-0372-2

[63] Ballari SA, Valenzuela AE, Nuñez MA. Interactions between wild boar and cattle in Patagonian temperate forest: Cattle impacts are worse when alone than with wild boar. Biological Invasions. 2020;**22**:1681-1689. DOI: 10.1007/s10530-020-02212-w

[64] Mayer H. Walbau auf soziologisch-ökologischer Grundlage. 4th ed. Germany: Gustav Fischer; 1992. p. 522

[65] Vaz AS, Kueffer C, Kull CA, Richardson DM, Vicente JR, Kühn I. Integrating ecosystem services and disservices: Insights from plant invasions. Ecosystem Services. 2017;**23**:94-107. DOI: 10.1016/j.ecoser.2016.11.017

[66] Bercê W, Bello C, Mendes CP, Vancine MH, Galetti M, Ballari SA. Invasive wild boar's distribution overlap with threatened native ungulate in Patagonia. Journal of Mammalogy. 2021;**102**:1298-1308. DOI: 10.1093/jmammal/gyab099

[67] Hodgkinson SM, Matus F, Lopez IF. Behavior of grazing European wild boar (Sus scrofa) in a semi-extensive production system. Ciencia e investigación agraria: revista latinoamericana de ciencias de la agricultura. 2013;**40**:193-199

Chapter 3

WSN System Warns Producer When Micro-Sprinklers Fail in Fruit Trees

Federico Hahn Schlam and Fermín Martínez Solís

Abstract

Salts in the irrigation water cause micro-sprinklers to clog. Farmers find it difficult to detect sprinkler clog due to the great number of trees grown in commercial orchards, causing a reduction in yield and timing problems. In this article, IoT can support farmers with daily soil moisture detection. A wireless sensor network, WSN was developed to warn the farmer from micro-sprinkling clogging. Trees were gathered into groups of 9 trees, where the central tree holds the master microcontroller and the other eight trees presented slave microcontrollers (nodes). The system uses BLE (Bluetooth Low Energy) to communicate between the master microcontroller by BLE. A second WSN using lasers was also tested but resulted to be a little more expensive. Soil moisture sensor performance against corrosion and current consumption was analyzed being the best sensors the V1.2 capacitance probe and the sprinkler-encoder one. When micro-sprinklers did not apply water to a tree, its number was transmitted via LoRa from the master to the producer's smartphone to warn him/her. A hexacopter was used to detect canopy stress from a height of 30 m, but only after 7 days of water removal did the NDVI indexes detect it.

Keywords: BLE, soil sensors, nuts, LoRa, WSN-smartphone, UAV, NDVI

1. Introduction

Mexico has planted areas exceeding 44,000 hectares of pecan trees (*Carya illinoinensis Koch*) in arid zones [1]. Production of pecan nuts in Mexico affects the United States' nuts prices as 95% of the world production of pecan nuts comes from North America (Mexico and the United States) [2]. Other countries producing pecan nuts are Australia, Israel, and Peru [3]. California plants 350,000 ha of almonds being 77% of the world's production [4, 5]. Almond trees require between 700 and 1500 mm of water to obtain good yields [6] so they need irrigation systems. A high-yield mature almond orchard in South Australia presented evapotranspiration (ETc) of 1430 mm [7]. ETc used annually by pecan trees in Egypt was 2100 mm [8]. Water use by almond trees depends on canopy cover and evaporative demand [9].

Better irrigation water use efficiency was obtained with pressurized systems than with furrow irrigation [10]. Drip irrigation saves 30% of the water applied to almond trees when compared with surface irrigation [4, 11]. Almond farms in Australia use high-efficiency drip systems using from 12 to 15 ($2\ l\ h^{-1}$) drippers per tree [12]. Pulsing irrigation and deficit irrigation increased water productivity. Less water productivity was increased using deficit irrigation. The modeled water uptake using the HYDRUS-2D software was higher than sap flow measurements before the almond harvest [12]. Sprinkler irrigation systems in pecan trees produced greater trunk growth diameter in contrast to other irrigation systems [13]. Wet soil volume from micro-sprinklers was important for almond yield [14]. Sustained deficit irrigation applied water at a given percentage of full ETc over the entire season and provided excellent yield [15].

Automatic irrigation systems require sensors to log irrigation variables [10]. Fruit trees are sensitive to water deficit, so irrigation evaluation becomes critical. Water stress monitoring during deficit irrigation allows saving water without decreasing fruit yield [16]. For fruit trees, monitoring the control of soil moisture is important to avoid the proliferation of fungi in the roots under high levels of humidity [17]. Tensiometers measure soil water potential and are useful for timing irrigations [18]. Continuous monitoring with dendrometers is time-consuming and problematic for automated irrigation [19]. Midday stem water potential (MSWP) has thus become the current standard irrigation management tool [20].

LoRaWAN radio signals within forests can be attenuated due to diffraction and scattering effects caused by tree obstacles [21–23]. Low frequencies provide good penetration within forests and present lower losses than high-frequency bands [24]. Tropical orchards can experiment with drastic weather conditions, being air temperature the variable that affects more LoRa transmission [25–27]. Air temperature decreases the received signal strength indicator (RSSI) variable [25]. Rain and relative humidity did not affect RSSI strongly [25, 28].

Precision agriculture in fruit crops has become fundamental for optimizing yields [29]. Internet of Things (IoT) has been used in recent years for plant growth monitoring [30]. Unmanned aerial vehicles (UAV) can observe trees that cover several hectares with multi-spectral cameras [31]. Normalized difference green near-infrared index (NDGNI) obtained with spectral cameras has also proved to be sensitive to water status [32]. Multispectral imagery remote sensing can monitor crop ET and crop water use throughout the growing season [33]. The normalized difference vegetation index provides a reliable estimate for leaf chlorophyll content and LAI [34]. The B5:B7 near-infrared ratio positively correlates with the moisture status of pecan orchards [35].

In this chapter, sensors detect when the micro-splinker system is not irrigating trees properly. The proposed remote monitoring network is able to: (1) monitor soil moisture in each tree of the orchard at sunset; (2) trigger an alarm if soil moisture was not detected; and (3) warn the producer by sending a short message system (SMS) to his/her smartphone. The WSN (wireless system network) for different moisture sensors and orchard densities are reviewed.

2. Walnut tree orchard

The experiment during the 2018–2019 season was carried out in the Western walnut-tree plantation at Delicias, Chihuahua, Mexico (28°11′35″ N, −105°28′18″ W, 1190 m). The 25-year-old orchard has a density of 10 × 10 trees per hectare with an average tree height of 7 m. Four micro-sprinklers were added per tree, together with

Figure 1.
Orchard monitoring provides a smartphone message to target the tree where irrigation problems were detected.

a water moisture sensor. Once the farmer receives a smartphone message (SMS) containing the tree number where irrigation failed, he/she drives to the tree and cleans the micro-sprinkler, letting it ready for the next irrigation cycle (**Figure 1**).

Tree number per hectare characterizes orchard density, which is important to increase yield. Almond tree size and height increase with age, and in order to obtain good yield, it is necessary to prune the orchard and let the radiation pass through the canopy. The reduced yield was obtained in almond low density (LD) farms with compost application [36]. Super high-density (SHD) orchards (over 2000 trees/ha) were studied since 2010 [37]. Kernel and almond in shell yield were 1.58 and 4.43 ton/ha, respectively [37]. The 25-year-old Western walnut-tree plantation in Delicias, Mexico, with a density of 10 × 10 trees per hectare, sprinkler irrigation, and a nitrogen application of 200 kg/ha produced 33 kg/tree [38]. Tree crowding reduces productivity due to excessive shading [39]. Tree management, including spraying and harvesting, becomes more difficult.

2.1 Soil and tree water content detector

Real-time monitoring of the whole orchard is impossible to carry out without sensors. Each tree presents a monitoring system consisting of moisture sensors, a microcontroller, and some of them a master microcontroller with a LoRa transmitter (TTGO LoRa32 OLED V2.1.6., Lilygo, China). A plastic box in each tree trunk covers a low power consumption CC2541 chip inside (**Figure 2**). The CC2541 built by Texas Instruments includes a Bluetooth Low Energy (BLE) platform, a built-in 8051 microcontroller, 256 KB of programmable memory, 8 KB of RAM, IO ports, and a 2.4-GHz RF transceiver.

2.1.1 Soil moisture detectors

Precise soil moisture measurements can be obtained with the gravimetric method, using tensiometers, neutron gauges, gamma-ray attenuation, and time-domain

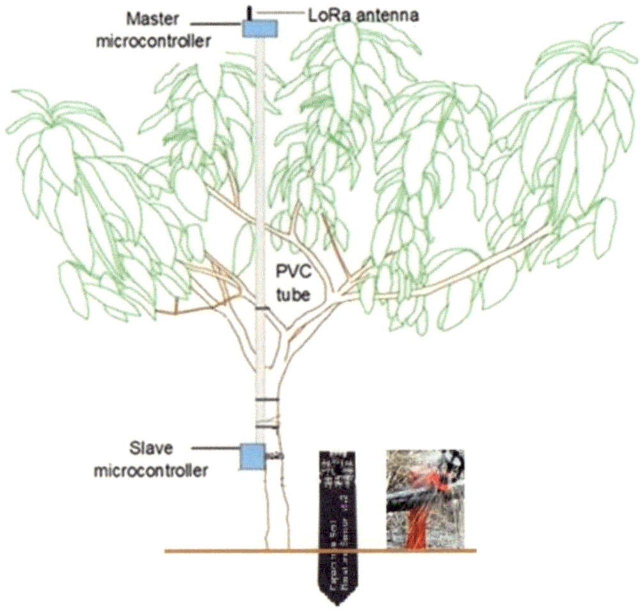

Figure 2.
Tree controller and transmitters.

reflectometry (TDR) [40]. Neutron probes, TDR, and gamma-ray sensors are highly accurate but extremely expensive. The neutron probe costs around $10,000 and estimates water in a larger area than other soil moisture sensors. TDR needs a very good data logger ($2000–3000) and each sensor cost is 100 US$.

Traditional farming depends on farmers' skills and experience. A transition to smart farming is noticed in advanced countries. Smart farming optimizes yields after measuring environmental and crop conditions. Soil moisture measured at multiple locations requires cheap sensors. Low-cost dielectric soil moisture sensors require complex processing circuits and are dependent on soil type and temperature [41]. Five different soil moisture sensors are compared in **Table 1** [17, 42], and prices are decreasing as smart technologies take over. The Decagon EC-5 sensor measures volumetric water content (VWC) after obtaining the dielectric constant by frequency

Model	Type	Sensit, %	VCD	Size mm	Cost $	Ref
MT-02A	Cap	3% 0–53 5% 53–100	3.6–30	149 × 45	79	[17]
YL-69	Res	0–95	3.3–5	60 × 30	2.6	[17]
Wat-200 s	Res	0–95	4–15	100 × 50	50	[42]
Theta Probe	Refr	xxxx	5–14	158 × 40	998	[43]
DecE-05	Diel	0–100	2.5–4	89 × 18	105	[42]
VH400	Diel	0–95	3.5–20	95 × 7	40	[42]

Table 1.
Comparison of relatively cheap soil moisture sensors.

domain technology [42]. As the sensor works above 70 MHz, its cost increases to 105 US$, due to all the electronics involved, **Table 1**. It works accurately in soilless media and under saline soils.

Remote dielectric and capacitive soil water content sensors are becoming attractive as the Delta-T ThetaProbe is being precise but very expensive [43]. The YL-69 soil moisture sensor [17] provides a variable electric signal when water is present between the electrodes and the soil. Ion movement in wet soil is higher than in dry soil. This sensor is equipped with an LM393 potentiometer and provides a digital output depending on the soil moisture [44]. Clay and loamy soil samples will provide different YL69 output values when the soil has a different moisture content [45]. Clay and loam soil with 20% moisture will obtain sensor measurements of 90 and 70%, respectively. In clay soils, sensor readings are linear only until 13% [45].

2.1.2 Cheap soil moisture measurements

In this experiment, several moisture sensors were used to compare efficiency, energy consumption, and cost.

- Capacitive V.1.2 sensor: This soil moisture sensor (Paialu, China) is buried at 6 cm and requires a DC voltage of 3.3 V (**Figure 3a**). The output is connected to a CC2541 microcontroller. This sensor is not affected by corrosion and its output should be close to 500 under dry conditions in clay and loam soils and 420 under wet conditions. This output is the one provided by the CC2541 microcontroller.

- Resistive YL-69 sensor: This sensor is smaller than the V1.2, so it is only buried at 5 cm. A potentiometric module YL-38 (**Figure 3b**) can be added to provide a digital output when it arrives at a given soil moisture threshold. The probe is calibrated using the air and water as dry and 100% humid values, respectively. Once buried in the soil (**Figure 3c**), it will not require further calibration.

- Wetness sensor: The PCB circuit board is composed of several cooper thin wires. Its resistance changes when water is sprayed (**Figure 4a**). The resistance reaches a maximum of 200 kΩ when the air is dry and 5 kΩ when air is moistened by sprinkler irrigation. Drops generally remain over the PCB board for some minutes.

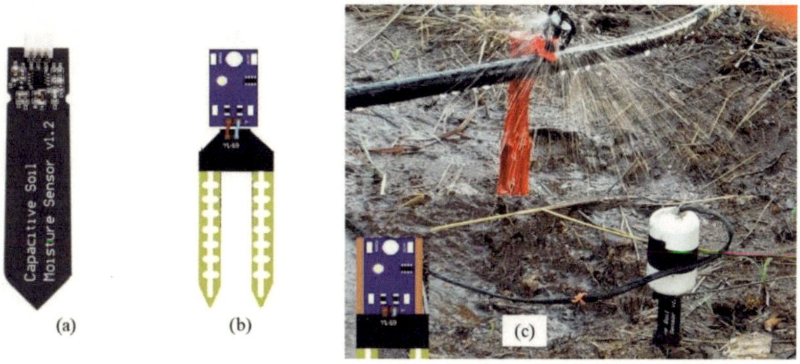

Figure 3.
Moisture (a) capacitive V1.2 probe, (b) YL-69 probe, and (c) V1.2 and YL-69 probes buried in soil.

Figure 4.
Sprayed water (a) to a wetness sensor, (b) to a dry solar cell, and (c) to forming drops in solar cell.

- Solar cell sensor: The 5-watt panel (**Figure 4b**) can be used to charge a capacitor. When the solar panel gets wet (**Figure 4c**), no charge will be supplied to the capacitor. Once the detection is carried out at sunset the capacitor is discharged.

- Sprinkler rotation sensor: This sensor detects whether water is being applied. As the micro-sprinkler starts operating, a circular washer on its top surface rotates. This washer presents one small orifice that rotates. A high-performance, low-cost, optical incremental encoder module (mod HEDS-973x, Avago Technologies, USA) counts the number of pulses and charges a capacitor during irrigation (**Figure 5a**). Ambient temperature can increase over 70°C and its voltage supply can work from 0.5 to 7 VDC.

- Air temperature and humidity sensor: The DHT11 is a low-cost digital sensor. It uses a capacitive humidity sensor but new data can only be acquired every 2 s. Humidity can be read from 20 to 80% and its energy consumption is of 2.5 mA per reading. It is recommended to hang it from the tree (**Figure 5b**) rather than from the hose, where drops begin to move toward the sensor and can damage it (**Figure 5c**).

2.2 Wireless system network

A wireless system network (WSN) monitored water application to each tree within one hectare. Every tree group consists of 9 trees (**Figure 6**). One hectare is regrouped into 12-groups, being water content detectors monitored at sunset. The central tree (A) from each group holds a master microcontroller module (TTGO LoRa32 OLED V2.1.6., Lilygo, China). This master module is fixed at a height of 5 m at the end of a hydraulic PVC tube (**Figure 2**). The remaining eight trees have CC2541-slave microcontrollers (SM).

The CC2541 uses BLE (Bluetooth Low Energy) for communication taking place at sunset. Once the slave microcontroller receives the packet from the slave transmitter, it responds. If there is no data to send, an empty packet is sent, finishing the BLE connection.

The first group array (**Figure 6a**) communicates via BLE sending a value at 18:00, just at sunset. A sensor detects if water did not reach a tree, so after CC2541 acquisition and processing, it will transmit the tree number to the following tree via BLE. The sequence starts with tree G, transmitting the tree number or 0 to tree E. Tree E detects whether it was irrigated, otherwise it transmits the tree number to tree H.

Figure 5.
Water measurement from (a) a micro-sprinkler with an optical sensor, (b) air moisture sensor DHT 11 hanging from the tree, and (c) DHT 11 hanging from lateral hose.

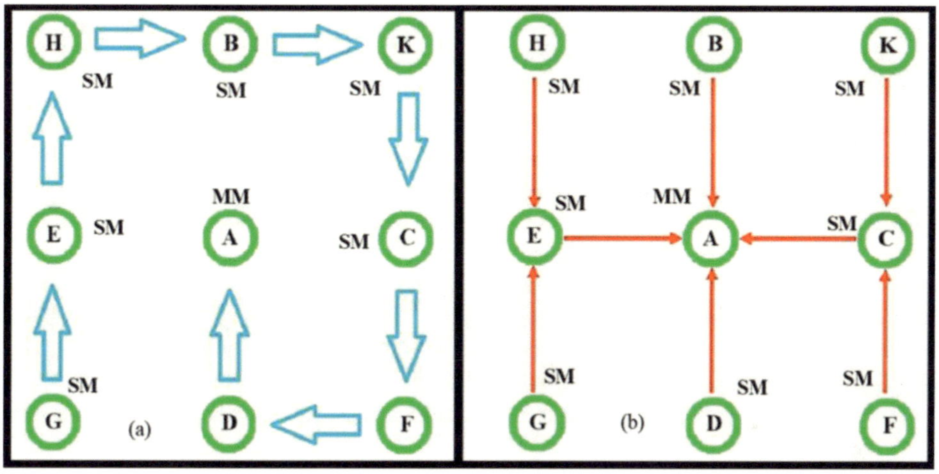

Figure 6.
Group of 9 trees using (a) BLE transmitters and (b) lasers for the orchard WSN.

This process continues until all the collateral trees send their information reaching the final value to the central tree. The values that will be sent to the producer's smartphone by the master microcontroller LoRa vary depending on the number of trees that were not irrigated. Trees are numbered from 1 to 108 in 1 ha, so the third group has trees numbered between 19 and 27. If G, E, and H failed due to a lateral problem, the data received by the master microcontroller will be 212,019 (**Figure 7a**). If only E and C trees were not watered, the data that will be arriving to the master is 2026 (**Figure 7a**). If all trees were watered, no data reaches the microcontroller.

Another group array topology (**Figure 6b**) substitutes the BLE signal with optical lasers. A CC2541-slave microcontroller (SM) was placed in each tree surrounding the central tree (A). The master microcontroller (MM) at tree A is the TTGO LoRa32 OLED V2.1.6. Each CC2541 reads the sensor signal, determines whether the tree was watered, and turns on the laser in case the tree soil is dry. A battery charged by a solar panel supplies the energy required to turn on the 650 nm–5 mW laser diode (D650-5I, USLASERS INC, USA). The phototransistor at the adjacent tree (10 m away) saturates after being lighted by a laser beam with a 14 mm diameter. A capacitor charges and

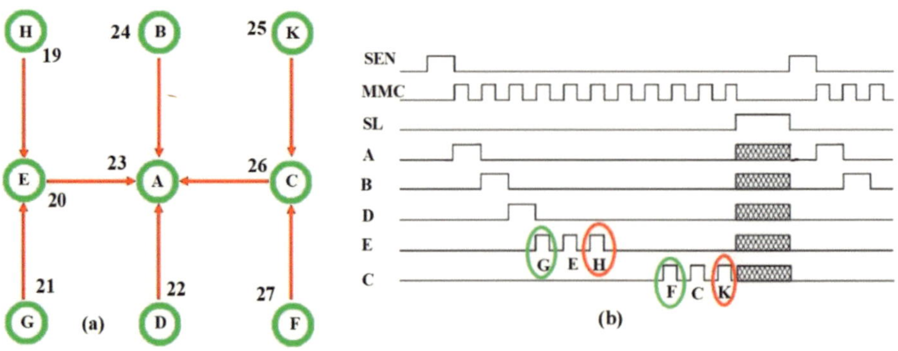

Figure 7.
Third group of 9 trees (a) using BLE lasers for WSN and (b) master microcontroller signals.

this voltage will be read by the CC2541. The phototransistor (BPW76A, Vishay, India) was chosen as it operates under a wide temperature range using an operating voltage up to 5 V and a collector current of 0.5 mA.

As lasers do not transmit electrical signals, the CC2541 turns on the laser providing a 1.5 s pulse. The CC2541 slave microcontroller on tree E acknowledges its own water status, before checking if the capacitors were charged by lasers from trees H and G. If no charge is detected and the tree was watered, the CC2541 goes to sleep (**Figurc 6b**). Tree A will hold 4 phototransistors to acknowledge the moisture status of the entire group. The master microcontroller timing signal explains its operation, being synchronized by the microcontroller clock (**Figure 7b**).

A light sensor indicates when sunset arrives, starting the operation routine. Once the MM is operating, it acquires the tree A sensor value first and waits 3 s for checking the capacitor status. It will charge through the phototransistor if tree B laser beam is on. If this is the case, it saves number 24 in a register (**Figure 7a**). The same operation takes place with tree D (number 22) 3 s later.

Information coming from trees E and C has to be decoded. It also contains information from trees H and G, and K and F. Laser from tree E sends 3-pulses of 1.5 s each, separated by OFF pulses of 1.5 s. The three pulses sent by the CC2541 controller are shown at the E timing signal (**Figure 7b**). Three pulses are sent, when trees G, E, and H were not watered. If only the first pulse appears (green pulse), tree G is not watered and the master microcontroller will convert it to number 21 (**Figure 7a**). The same operation takes place with tree C. After checking all the trees, the MM generates a value that is transmitted through LoRa, and the tree enters sleep mode until the next day.

2.2.1 Master LoRa communication

LoRa provides a low-cost communication system that does not require a license. TTGO LoRa32 OLED V2.1.6. modules can communicate through BLE and LoRa. The LoRa was programmed at 915 MHz, with a power transmission of 20 dBm, a spreading factor (SF) of 12 and a bandwidth of 125 kHz. The maximum packet size for a 9-tree group would be 18 bytes if all the trees were not watered at the same time.

Another TTGO LoRa32 OLED V2.1.6 module at the roof of the farmer's house works as a general receiver from the entire plantation and presents a 3 dbi 915 MHz whip antenna. In addition, a SIM Card is incorporated so that the tree number with water shortage could be transferred to the farmer's smartphone by SMS.

Transmission between the 12 central tree modules within a hectare and the end TTGO LoRa32 OLED V2.1.6 module at the orchard house was synchronized to avoid crossover between SMS messages. The time delay for this transmission takes 13 s. The 3 central trees further away transmitted their data 1–2-min after sunset. LoRa from the following 3 groups transmitted data 3–4 min after sunset and so on.

2.3 Optical remote sensing measurements

In situ soil moisture monitoring of crop water stress is time-consuming, and assumes uniform plant density and transpiration rate [46]. Vis–NIR (400–1100 nm) can provide measurements of soil moisture [47]. Hydration, hygroscopic, and free water are present in the soil, being the free water inside soil pores. Spectral absorption peaks for hygroscopic and free water occur at 1900 nm but cannot be measured with a cheap VIS–NIR spectrometer. Vis–NIR optical soil moisture measurements require large local calibration data sets limiting their use [48]. Remote sensing detects vegetation stress in trees [49, 50], but satellites provide low-resolution images with a limited field of view. Nano-satellites increase spatial resolution and image timing [50]. NDVI (Normalized Difference Vegetation Index) [51] and crop water stress index [49, 52] have been correlated against canopy temperature and tree stress. NDVI provides useful information on canopy structure and drought stress [53].

2.3.1 UAV selection and flight programming

A six-wing DJI S900 (DJI Co., Ltd. Shenzhen, China) with a load capacity of 5 kg was selected to take tree images (**Figure 8a**). Time of flight at midday took 18 min and was done at a height of 30 m at a speed of 4 km h^{-1}. Multi-rotor drones used to study pistachio, almond and walnut plantations collect RGB, and multi spectral images [54]. The hexacopter was equipped with a Parrot Sequoia multispectral camera (SenseFly, Inc., Switzerland). The Parrot equipment presents 2 sensors: a

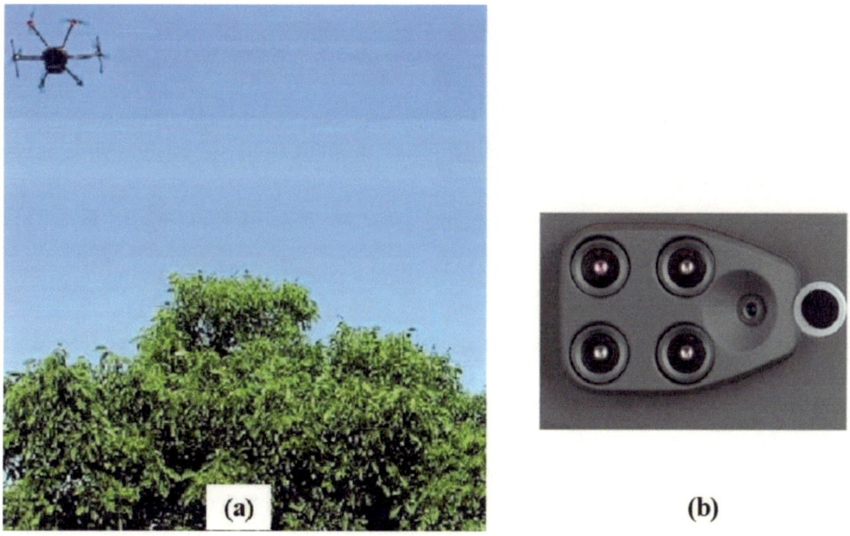

Figure 8.
Hexacopter (a) flying over tree and (b) parrot camera.

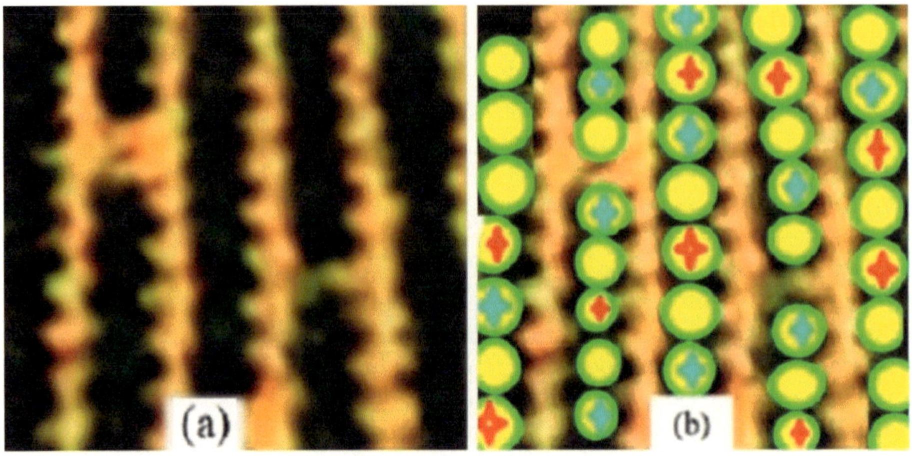

Figure 9.
Image taken by the hexacopter (a) in RGB and (b) showing trees with sprinklers removed.

multispectral camera and a light sensor to monitor sunlight intensity. The multispectral sensor has four (1.2 megapixels) monochrome cameras: green, red, red edge, near-infrared, and 16 MP RGB [55]. Their RGB and multispectral spatial resolution were of 1.2 cm/px and 3.8 cm/x, respectively [56]. The hexacopter took aerial images of the plantation on the day when some micro-sprinklers were removed as well as 4 and 7 days after (**Figure 9a**). Aerial images were taken every 2 s, having an overlap of 80%. Some trees had all their 4 micro-sprinklers removed (red cross), meanwhile those trees with a blue cross had only one sprinkler removed (**Figure 9b**). The experiment was repeated 3 times. After the UAV returned, images were exported for processing with Pix4D software (Pix4D v.3.1., Switzerland).

3. Results and discussion

The results obtained can determine the best cheap sensors for sensing if the micro-sprinklers applied water in the orchard, being a useful precise agriculture application. Transmission effects due to laser, BLE, and LoRa are all considered. UAV data were analyzed to obtain the NDVI index to determine if sprinkler application can be detected remotely.

3.1 Sensor experimental comparison

A general comparison of the sensors used within the plantation is shown in **Table 2**. The first two sensors (capacitive V1.2 and YL-69 probes) were tested in sixty trees studying their performance during the month of April (**Figure 10a** and **b**). When soil moisture increases, the conductivity of soil rises and the YL-69 values vary between 400 and 650 (**Figure 10b**). These values result from the 10-bit ADC (analog-digital conversion) carried out by the CC2541 microcontroller. The sensor analog output provides moisture levels that will be converted to a digital signal that varies between 0 and 1023.

Dry clay and loam soil monitored with YL-69 probes presented average ADC values of 503 and 640, respectively. Their standard deviations were 28 and 22,

Model	Sensitivity %	Efficiency	Current, mA	Wireless system	Cost $
V 1.2 sensor	3–95	100	5	A, B	2.8
YL-69 sensor	0–95	100	5	A	2.65
Wetness PHY31	0–95	78	3	B	65
NORPS-12	0–95	82; 35	0.47	B	1.8
Sprinkler encoder	0–100	100	5	A, B	7
DH11 sensor	0–95	100	2.5	B	2.4

*A: BLE system (***Figure 6a***); B: laser system (***Figure 6b***).*

Table 2.
Comparison of sensors for detecting micro sprinkler application.

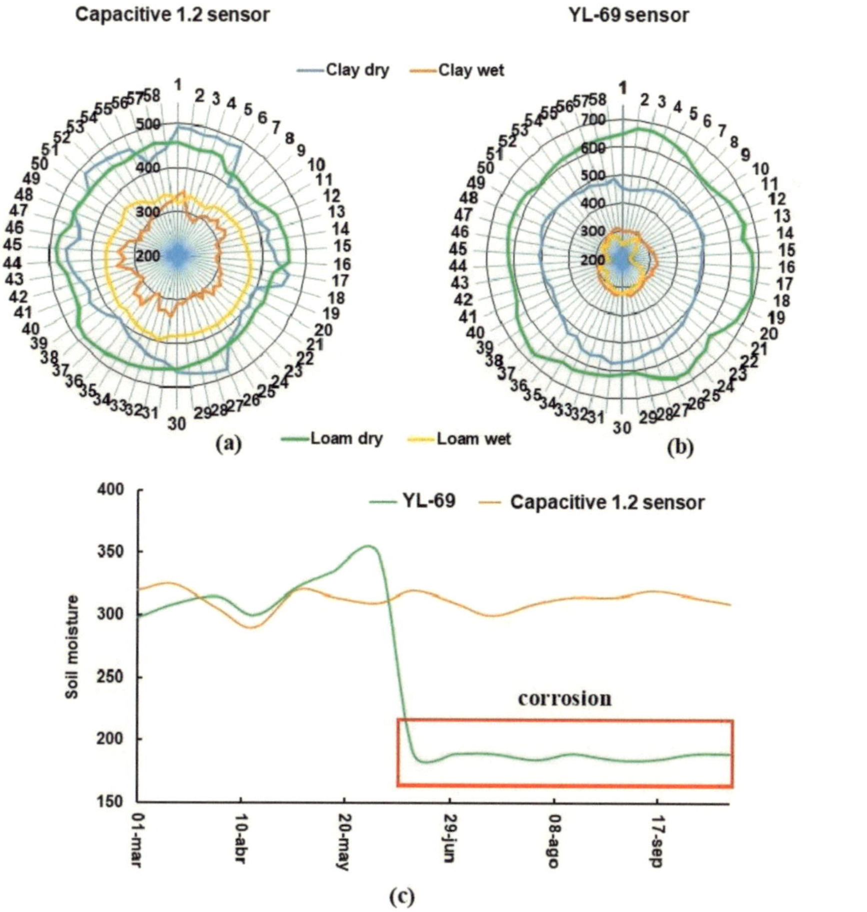

Figure 10.
Measurements of (a) 60 capacitive 1.2 sensors, (b) 60 YL-69 sensors for dry and wet soils, and (c) both probes during a period of 6 months.

respectively. After micro-sprinkler irrigation, soil moisture increased at least by 50%. YL-69 average measurements in clay and loam decreased to 307 and 276, respectively. Their standard deviations were 15.21 and 26.9, respectively. The variable resistance

YL-69 probe's maximum wet value for both soils was 338 and the minimum dry value was 449 in clay soil, being easy to discriminate.

A similar study carried out by Adla et al. [57] provided excellent results for monitoring soil moisture in all kinds of soils including those with sand. Soil resistivity is influenced by salt concentration in water [58] or by salts added as fertilizers. A sensor was left in the soil nearby the tree and it did not respond after 3 months (**Figure 10c**). YL-69 probes coated with nickel plating rusted after 4 days requiring replacement [59].

The capacitive 1.2 sensor is a cheaper version of the SM100 Soil Moisture sensor built by Spectrum Technologies [57]. In dry clay and loam soil, average moisture was similar being 442 and 455, respectively. Wet average soil clay and loam moisture values were 316 and 367, respectively. Their standard deviations were 16.4 and 17.3, respectively. Minimum dry clay soil measurement was 396, meanwhile maximum wet soil measurement in clay was 361. A 100% discrimination was obtained as noted by the blue and orange lines of **Figure 10a**. Dry and wet soil was also perfectly discriminated in loam soils (**Figure 10a**). Sensors made of a corrosion-resistant material monitored wet clay soil for 8 months, varying between 280 and 330 (**Figure 10c**). These sensors supplied by AC (alternating current) do not rust, resulting in longer operating life.

The solar cell sensor (**Table 2**) provides an extremely small current due to the short operation time and poor illumination intensity at dusk. The solar cell was substituted by a light-dependent resistance (LDR) that measures illumination. The selected LDR (mod. NORPS-12, Silonex, the UK) is encapsulated with a humidity-resistant coating. The resistance of the LDR is 6 kΩ when it is illuminated by a LED and increases to 1 MΩ during dusk. A battery of 3.3 V supplies voltage to a serial divider formed by a 1 kΩ resistance and the LDR. As the sprinkler deposits water drops over the LDR surface, LED incoming radiation gets scattered within the water. A voltage of 2.83 V is obtained at the divider pin when the LED illuminates the dry LDR surface. With water drops over the LDR surface, the voltage increases to 3.29 V at the divider. The experiment was carried out with one hundred sensors, its efficiency being 83% and 35%, during calm and windy days, respectively. The wind does not allow the drops to stay over the LDR. On calm days, erroneous LDR measurements were found when the sensor slope increased as water drops fell to the soil.

Leaf wetness duration (LWD) sensors provide important information for the prediction of plant disease [60]. Capacitive leaf wetness sensors have been developed for IoT applications with Arduino but their cost is relatively high (48 US$). When it operates at a frequency of 1 kHz, the sensor responds in 5 s [61]. Leaf wetness sensors (LWS) built, nowadays, on flexible substrates can detect whether the leaf canopy is dry or wet [61]. As water appears over the surface substrate, the dielectric increases.

The Phytos 31 LWS (4 × 6 cm) produces a variable linear voltage according to the surface covered by the water. The sensor was tested around the plantation in different trees to study its efficiency for detecting sprinkler spraying. With the dry sensor, the CC2541 should acquire a value beneath 0.4 mV and when 40% of the LWS area is humid, the voltage provided by the sensor should be over 0.48 mV. Each tree having an LWS was sensed fifteen times during the irrigation period and the average value was saved. After 15 min, the probability to get more drops over the sensor was higher. The efficiency for detecting sprinkler operation was 78%, being dependent on wind, sensor slope, and proper positioning of the irrigation sprinkler. Grasshoppers and spiders were also attracted by the Phytos 31 LWS.

The most innovative system to detect water leaving the sprinkler is to use a C-shaped emitter/detector module. Coupled with a washer fixed to the sprinkler, it translates rotary movement into a digital output. This was achieved with a HEDS-973X circuit from Avago Technologies. When water is supplied, the sprinkler turns the washer that has a small orifice in it. As light from the emitter diode passes through the orifice, it generates a pulse. The pulse interrupts the CC2541, and after 10 interruptions it will recognize that water was applied. This system worked with 100% accuracy and was not dependent on environmental conditions.

DHT 11 showed extreme variations during the first day of analysis and most sensors got damaged after getting wet for 3 days with saturated water. They were substituted in some trees by the DHT 22 that theoretically withstands 100% RH but also failed after 1 week. The DTH40 (Sensirion, Switzerland) can measure from 0 to 100% RH and it can do it every 8 s. Its energy consumption is 28 μA and its cost is 2.87 US$, but still has to be tested in the field.

3.2 Best sensors and current consumption

Section 3.1 compared each sensor and only the capacitance V1.2 moisture sensor and the C-emitter/detector adapted to the sprinkler passed the requirements for great extensions due to their reduced cost and efficiency. These sensors work properly under natural environmental conditions and present high detection accuracy. Sensor current consumption depends on current demand and time of operation per day. For the capacitance sensor, it will take 5 s to measure the soil moisture content. In this interval, several samples are obtained and averaged, consuming 25 mAs per day. The sensor adapted to the sprinkler takes only 1 s to measure. Once this time is multiplied by the 5 mA current used (**Table 3**), it provides the current consumption per day (5 mAs). The costs per hectare for both type of detectors change, needing 400 sensors each. With 400 capacitive V1.2 probes connected to 100-CC2541 controllers, the cost will be 1760 US$ per hectare. Four hundred sprinkler-encoder sensors connected to one hundred CC2541 microcontrollers inside 1 ha will become more expensive: 4000 US$.

3.3 WSN transmission

Energy consumption of nodes within the BLE (A) and BLE-laser WSN systems are compared in **Table 3**. Current consumption by each node (tree) depends on the sensor used, CC2541 acquisition and processing, and finally transmission current. The CC2541 microcontroller employs 3 and 18.2 mA for each acquisition and transmission/reception, respectively. For the BLE WSN (A) with the soil capacitor sensor V1.2, it will consume (25 + 3 + 18.2 = 46.2) mA per day. In the case of the laser WSN, each laser needs 20 mA per second. The CC2541 of trees E and C (**Figure 7a**) turn on the laser 3 times

Sensor/WSN	Sensor, mA	Acq and Proc, mA	Trans/receipt, mA	Total energy, mA day^{-1}
V 1.2/A	25	3	18.2	46.2
V 1.2/B	25	3	90	118
Sprinkler A	5	1.5	18.2	24.7
Sprinkler B	5	1.5	90	96.5

*A: BLE system (**Figure 6a**); B: laser system (**Figure 6b**); Acq and Proc: acquisition and processing.*

Table 3.
Comparison of sensor nodes for detecting micro sprinkler application.

(1.5 s pulses) when no water is applied by the micro-sprinklers. If 20 mA is supplied to the laser, it will consume 90 mA per day in 4.5 s. All the other trees will consume 30 mA per day during laser communication if water fails. The sprinkler sensor provides pulses to interrupt the CC2541, so lower power consumption of 1.5 mA is required. The node that performs better from an energy point of view is the one having the sprinkler sensor and transmits the data through CC2541 BLE controller (**Table 3**); it only employs 24.7 mA per day. The phototransistors charge the capacitors in two seconds, so the energy employed was relatively low and was not included in **Table 3**.

3.3.1 Node costs

The costs per hectare for both sensors, together with its WSN network, change. The cheaper system will use V1.2 probes with the BLE CC2541 communication net (1760 US). If ten TTGO LoRa32 OLED V2.1.6 modules are needed to transmit data from 1 ha to the farmer's house, the final cost for all the intelligent systems will be 2200 US$. In the case of using the laser WSN (option B) with the V1.2 probes and the LoRa modules, the total cost increases to 2950 US$. Laser WSN using sprinkler-encoder sensors and LoRa will increase the cost to 5100 US$, meanwhile the sprinkler encoder sensors plus BLE WSN and LoRa will cost 4350 US$.

3.4 UAV images and NDVI

UAV images were taken at midday on April 3rd, 6th, and 9th at three different zones of the plantation. Many trees in the multispectral images (19 + 10) had null NDVI changes after 1 week. In the 19 trees, irrigation was maintained and the other 10 marked with a blue cross (**Figure 9b**) had only one sprinkler removed. After 1 week, those trees without irrigation showed NDVI changes (**Figure 11a** and **b**). Trees grown in loam soil are marked with circles filled with white, being more stressed than those grown over clay soil (red circles). After calculating the NDVI index, trees in loam soil decrease their value from 0.85 to 0.82, as schemed by the blue columns (**Figure 11b**), even when only one sprinkler was removed. The NDVI index of trees standing on clay soil had no variation during the first 4 days when the four sprinklers were removed as shown by the brown column in **Figure 11b**. After 7 days, NDVI index of trees grown on clay soil decreased to 0.81.

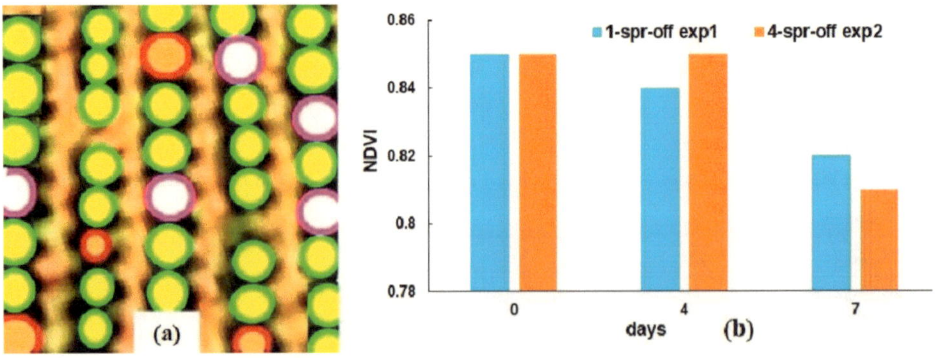

Figure 11.
UAV tree (a) having some stress, and (b) NDVI during the first, fourth, and seventh day.

4. Conclusions

Precision agriculture is a technological reality in orchards. In our study, soil sensors resulted in a better option than UAV imaging. UAV platforms require more research and different bands to detect how micro-sprinkler irrigation is working in real-time. Imaging analysis is much more complicated than simple embedded systems and producers. Producers find UAV technology complicated and environmental conditions make them much more difficult to use than simple embedded systems. Imaging analysis obtained by the camera is challenging for them.

Soil moisture sensors in a walnut orchard were tested; some of them as the DHT 11 and YL-69 failed after 1 week of operation. When drops of saturated water pour over the DHT 11 sensor, it failed after 3 days, meanwhile the YL-69 got rusty after applying a direct current voltage to it. The capacitive 1.2 probe covered with a corrosion-resistant material proved to measure properly during several seasons and under clay and loam soil. The V1.2 probe is very cheap and energy-efficient. The sprinkler-decoder sensor was the more efficient sensor based on current consumption, but its price increased by a factor of 2.5. Wetness sensors are being developed with flexible substrates to detect dry and wet conditions but are still very expensive.

WSN groups of nine trees were formed by hectare within the orchard; each group transmitted its information through a LoRa module. The BLE node proved much more economical than the laser system and both operate properly. LoRa modules transmit information toward the farmers' house sending the tree number where the micro-sprinkler failed. LoRa is able to transmit at distances of 500 m even when the temperature ranges between 30 and 35°C. The WSN system can be used to transmit other information from the plantation using a similar version of tree grouping.

Acknowledgements

We would like to acknowledge Eng. Angel Hernandez, Biol. Ruth Perez for the discussions and to the University Autonoma Chapingo for supporting and funding this project 19008-DTT-65.

Author details

Federico Hahn Schlam[1*] and Fermín Martínez Solís[2]

1 Irrigation Department, Universidad Autónoma Chapingo, Carretera México-Texcoco, Chapingo, Mexico

2 Engineering Department, Universidad Juárez Autónoma de Tabasco, Tabasco, Mexico

*Address all correspondence to: fedhahn@gmail.com

References

[1] Martinez DG, Núñez MH. Current status of pecan production in Mexico and future outlook. In: Proceedings of the 77 Annual Convention of Oklahoma Pecan Growers Association. 2007. 434 pp

[2] Castillo IO, Sangerman DM, Hernández MF, Vázquez CV, Robles MA. Production and marketing of pecan nuts (*Carya illinoensis Koch*) in northern Coahuila, Mexico. Revista Mexicana de Ciencias Agrícolas. 2013;**4**(3):461-476. Corpus ID: 128129723

[3] Barrett D, Somogyi L, Ramaswamy H. Processing Fruits. 2nd ed. Boca Raton, Florida, USA: CRC Press LLC. ISBN: 0-8943-1478-X. 841 pp

[4] Collin G, Caron J, Létourneau G, Gallichand J. Yield and water use in almond under deficit irrigation. Agronomy Journal. 2019;**111**:1381-1391. DOI: 10.2134/agronj2018.03.0183

[5] INC. International Nut & Dried Fruit. Nuts & Dried Fruits Statistical Yearbook 2019/2020. 2019. Available from: https://www.nutfruit.org/industry/statistics

[6] Micke WC, Kester DE. Almond growing in California. Acta Horticulturae. 1998;**470**:21-28. DOI: 10.17660/ActaHortic.1998.470.1

[7] Stevens RM, Ewenz CM, Grigson G, Conner SM. Water use by an irrigated almond orchard. Irrigation Science. 2012;**30**(3):189-200. DOI: 10.1007/s00271-011-0270-8

[8] El-Soda AS, Hend IA, El-Hisseny AM, Eid TA. Effect of irrigation levels on growth and productivity of pecan trees. American-Eurasian Journal of Agricultural & Environmental Sciences. 2021;**21**(1):60-71. DOI: 10.5829/idosi.aejaes.2021.60.71

[9] Sánchez JM, Simón L, González-Piqueras J, Montoya F, López-Urrea R. Monitoring crop evapotranspiration and transpiration/evaporation partitioning in a drip-irrigated young almond orchard applying a two-source surface energy balance model. Water. 2021;**13**:2073. DOI: 10.3390/w13152073

[10] Holzapfel EA, Pannunzio A, Lorite I, Oliveira AS, Farkas I. Design and management of irrigation systems. Chilean Journal of Agricultural Research. 2009;**69**:17-25. DOI: 10.4067/S0718-58392009000500003

[11] Ibragimov N, Evett SR, Esanbekov Y, Kamilov BS, Mirzaev L, Lamers JPA. Water use efficiency of irrigated cotton in Uzbekistan under drip and furrow irrigation. Agricultural Water Management. 2007;**90**:112-120. DOI: 10.1016/j.agwat.2007.01.016

[12] Phogat V, Skewes M, Mahadevan M, Cox JW. Seasonal simulation of water and salinity dynamics under different irrigation applications of almond in pulsed and continuous mode. In: 4th International Conference on HYDRUS Software Applications to Subsurface Flow and Contaminant Transport Problems. March 21-23, 2013; Prague, Czech Republic. ISBN: 978-80-213-2368-1

[13] ShalekBriski A, Brorsen BW, Biermacher JT, Rohla CT, Chaney W. Effect of irrigation method on tree growth, foliar nutrient levels, and nut characteristics of young pecan trees in the southern Great Plains. HortTechnology. 2019;**29**(2):109-113. DOI: 10.21273/HORTTECH04162-18

[14] Girona J, Mata M, Marsal J. Regulated deficit irrigation during the

kernel-filling period and optimal irrigation rates in almond. Agricultural Water Management. 2005;**75**:152-167. DOI: 10.1016/j.agwat.2004.12.008

[15] Goldhamer DA, Viveros M, Salinas M. Regulated deficit irrigation in almonds: Effects of variations in applied water and stress timing on yield and yield components. Irrigation Science. 2006;**24**:101-114. DOI: 10.1007/s00271-005-0014-8

[16] Zarco-Tejada PJ, Gonzalez-Dugo V, Berni J. Fluorescence, temperature and narrow-band indices acquired from a UAV platform for water stress detection using a micro-hyperspectral imager and a thermal camera. Remote Sensing of Environment. 2012;**117**:322-337. DOI: 10.1016/j.rse. 2011.10.007

[17] Lloret J, Sendra S, Garcia L, Jimenez JM. A wireless sensor network deployment for soil moisture monitoring in precision agriculture. Sensors. 2012;**21**:7243. DOI: 10.3390/s21217243

[18] Shock CC, Wang FY. Soil water tension, a powerful measurement for productivity and stewardship. HortScience. 2011;**46**(2):178-185. DOI: 10.21273/HORTSCI.46.2.178

[19] Jones HG. Irrigation scheduling: Advantages and pitfalls of plant based methods. Journal of Experimental Botany. 2004;**55**(407):2427-2436. DOI: 10.1093/jxb/ erh213

[20] Fereres E, Goldhamer DA. Suitability of stem diameter variations and water potential as indicators for irrigation scheduling of almond trees. Journal of Horticultural Science and Biotechnology. 2003;**78**(2):139-144. DOI: 10.1080/14620316.2003.11511596

[21] Azevedo JA, Santos FE. An empirical propagation model for forest environments at tree trunk level. IEEE Transactions on Antennas and Propagation. 2011;**59**(6):2357-2367. DOI: 10.1109/TAP.2011.2143664

[22] Meng Y, Lee YH, Ng BC. Empirical near ground path loss modeling in a forest at VHF and UHF bands. IEEE Transactions on Antennas and Propagation. 2009;**57**(5):1461-1468. DOI: 10.1109/TAP.2009.2016703

[23] Wiyadi E, Setiadi R, Umar L. Effect of vegetation profile and air data rate on packet loss performance of LoRa E32-30dBm 433 MHz as a wireless data transmission. Journal of Physics: Conference Series. 2020;**1655**:012015. DOI: 10.1088/1742-6596/1655/1/012015

[24] Rizman ZI, Jusoff K, Rais SS, Bakar HH, Nair G, Ho YK. Microwave signal propagation on oil palm trees: Measurements and analysis. International Journal on Smart Sensing and Intelligent Systems. 2011;**4**(3):388-401. DOI: 10.21307/IJSSIS-2017-446

[25] Elijah O, Abdul Rahim S, Sittakul V, Al-Samman A, Cheffena M, Din J, et al. Effect of weather condition on LoRa IoT communication technology in a tropical region: Malaysia. IEEE Access. 2021;**9**:72835-72843. DOI: 10.1109/ACCESS.2021.3080317

[26] Bezerra NS, Åhlund C, Saguna S, Sousa VA. Temperature impact in LoRaWAN—A case study in northern Sweden. Sensors. 2019;**19**(20):4414. DOI: 10.3390/s19204414

[27] Anzum R. Factors that affect LoRa propagation in foliage medium. Procedia Computer Science. 2021;**194**:149-155. DOI: 10.1016/j.procs.2021.10.068

[28] Fang SH, Cheng YC, Chien YR. Exploiting sensed radio strength and precipitation for improved distance

estimation. IEEE Sensors Journal. 2018;**18**(16):6863-6873. DOI: 10.1109/JSEN.2018.2851149

[29] Zhang C, Valente J, Kooistra L, Guo L, Wang W. Orchard management with small unmanned aerial vehicles: A survey of sensing and analysis approaches. Precision Agriculture. 2021;**22**:2007-2052. DOI: 10.1007/s11119-021-09813-y

[30] Wu Y, Guo G, Tian G, Liu W. A Model with leaf area index and trunk diameter for LoRaWAN radio propagation in Eastern China Mixed Forest. Journal of Sensors. 2020;Article ID 2687148. pp. 1-16. DOI: 10.1155/2020/2687148

[31] Bendig J, Bolten A, Bareth G. Introducing a low-cost mini-UAV for thermal and multispectral imaging. The International Archives of the Photogrammetry, Remote Sensing and Spatial Information Sciences. 2012;**39**:345-349. DOI: 10.5194/isprsarchives-XXXIX-B1-345-2012

[32] Bulanon DM, Lonai J, Skovgard H, Fallahi E. Evaluation of different irrigation methods for an apple orchard using an aerial imaging system. ISPRS International Journal of Geo-Information. 2016;**5**(6):79. DOI: 10.3390/ijgi5060079

[33] Calera A, Campos I, Osann A, D'Urso G, Menenti M. Remote sensing for crop water management: From ET modelling to services for the end users. Sensors. 2017;**17**:1104. DOI: 10.3390/s17051104

[34] Houborg R, Boegh E. Mapping leaf chlorophyll and leaf area index using inverse and forward canopy reflectance modeling and SPOT reflectance data. Remote Sensing of Environment. 2008;**112**:186-202. DOI: 10.1016/J.RSE.2007.04.012

[35] Othman Y, Steele C, VanLeeuwen D, Heerema R, Bawazir A, St. Hilaire R. Remote sensing used to detect moisture status of pecan orchards grown in a desert environment. International Journal of Remote Sensing. 2014;**35**:949-966. DOI: 10.1080/01431161.2013.873834

[36] De Leijster V, Verburg RW, Santos MJ, Wassen MJ, Martínez-Mena M, de Vente J, et al. Almond farm profitability under agroecological management in South-Eastern Spain: Accounting for externalities and opportunity costs. Agricultural Systems. 2020;**183**:102878. DOI: 10.1016/j.agsy.2020.102878

[37] Casanova-Gascón J, Figueras-Panillo M, Iglesias-Castellarnau I, Martín-Ramos P. Comparison of SHD and open-center training systems in almond tree orchards cv. 'Soleta'. Agronomy. 2019;**9**:874. DOI: 10.3390/agronomy9120874

[38] Tarango-Rivero H. Manejo del nogal pecanero con base en su fenología. [Pecan orchard management based in its phenology]. Folleto Técnico 24, Campo Experimental Delicias. 3a ed. Chihuahua: INIFAP; 2012. 39 pp. ISBN: 978-607-425-769-4

[39] Andersen PC, Crocker TE. The pecan tree: HS984/HS229, 5/2004. 1st ed. Florida Cooperative Extension Service, Institute of Food and Agricultural Sciences, University of Florida; 2004. HS984. DOI: 10.32473/edis-hs229-2004

[40] Yu L, Tao S, Ren Y, Gao W, Liu X, Hu Y, et al. Comprehensive evaluation of soil moisture sensing technology applications based on analytic hierarchy process and Delphi. Agriculture. 2021;**11**:1116. DOI: 10.3390/agriculture11111116

[41] Kojima Y, Shigeta R, Miyamoto N, Shirahama Y, Nishioka K, Mizoguchi M,

et al. Low-cost soil moisture profile probe using thin-film capacitors and a capacitive touch sensor. Sensors. 2016;**16**(8):1292. DOI: 10.3390/s16081292

[42] Payero JO, Qiao X, Khalilian A, Mirzakhani-Nafchi A, Davis R. Evaluating the effect of soil texture on the response of three types of sensors used to monitor soil water status. Journal of Water Resource and Protection. 2017;**9**:566-577. DOI: 10.4236/jwarp.2017.96037

[43] Huang Q, Akinremi OO, Sri Rajan R, Bullock P. Laboratory and field evaluation of five soil water sensors. Canadian Journal of Soil Science. 2004;**84**:431-438. DOI: 10.4141/S03-097

[44] Fauziyah M, Safitri H, Dewatama D, Aulianta E. Conditioning of temperature and soil moisture in chrysanthemum cut flowers greenhouse prototype based on internet of things (IoT). ELKHA: Journal Teknik Elektro. 2021;**13**(1):25-32. DOI: 10.26418/elkha.v13i1.43078

[45] Sudha LK, Sinha A. Effect of moisture content in the sandy, clay and loamy soil using humidity sensor YL 69. International Journal of Science and Research (IJSR). 2017;**6**(5):107-109. Corpus ID: 53474495

[46] Ihuoma SO, Madramootoo CA. Recent advances in crop water stress detection. Computers and Electronics in Agriculture. 2017;**141**:267-275. DOI: 10.1016/j.compag.2017.07.026

[47] Marakkala Manage LP, Humlekrog Greve M, Knadel M, Moldrup P, De Jonge LW, Katuwal S. Visible-near-infrared spectroscopy prediction of soil characteristics as affected by soil-water content. Soil Science Society of America Journal. 2018;**82**:1333-1346. DOI: 10.2136/sssaj2018.01.0052

[48] Hardie M. Review of novel and emerging proximal soil moisture sensors for use in agriculture. Sensors. 2020;**20**:6934. DOI: 10.3390/s20236934

[49] Bellvert J, Adeline K, Baram S, Pierce L, Sanden BL, Smart DR. Monitoring crop evapotranspiration and crop coefficients over an almond and pistachio orchard throughout remote sensing. Remote Sensing. 2018;**10**:2001. DOI: 10.3390/rs10122001

[50] Machwitz M, Pieruschka R, Berger K, Schlerf M, Aasen H, Fahrner S, et al. Bridging the gap between remote sensing and plant phenotyping—Challenges and opportunities for the next generation of sustainable agriculture. Frontiers in Plant Science. 2021;**12**:749374. DOI: 10.3389/fpls.2021.749374

[51] Rocha J, Perdiao A, Melo R, Henriques C. Remote sensing based crop coefficients for water management in agriculture. In: Curkovic S, editor. Sustainable Development—Authoritative and Leading Edge Content for Environmental Management. London, UK: IntechOpen Limited; 2012. pp. 167-191. Chapter 8

[52] González-Dugo V, Zarco-Tejada PJ, Fereres E. Applicability and limitations of using the crop water stress index as an indicator of water deficits in citrus orchards. Agricultural and Forest Meteorology. 2014;**198-199**:94-104. DOI: 10.1016/j.agrformet.2014.08.003

[53] Wong CYS, Bambach NE, Alsina MM, McElrone AJ, Jones T, Buckley TN, et al. Detecting short-term stress and recovery events in a vineyard using tower-based remote sensing of photochemical reflectance index (PRI). Irrigation Science. 2022;**40**(3). DOI: 10.1007/s00271-022-00777-z

[54] Jafarbiglu H, Pourreza A. A comprehensive review of remote sensing

platforms, sensors, and applications in nut crops. Computers and Electronics in Agriculture. 2022;**197**:106844. DOI: 10.1016/j.compag.2022.106844

[55] Teixeira AA, Mendes J, Cláudio W, Bredemeier C, Negreiros M, Aquino R. Evaluation of the radiometric accuracy of images obtained by a sequoia multispectral camera. Engenharia Agrícola. 2020;**40**(6):759-768. DOI: 10.1590/1809-4430-Eng.Agric.v40n6p759-768/2020

[56] Candiago S, Remondino F, De Giglio M, Dubbini M, Gattelli M. Evaluating multispectral images and vegetation indices for precision farming applications from UAV images. Remote Sensing. 2015;7:4026-4047. DOI: 10.3390/rs70404026

[57] Adla S, Rai NK, Karumanchi SH, Tripathi S, Disse M, Pande S. Laboratory calibration and performance evaluation of low-cost capacitive and very low-cost resistive soil moisture sensors. Sensors. 2020;**20**:363. DOI: 10.3390/s20020363

[58] Bouyoucos GJ, Mick AH. A comparison of electric resistance units for making a continuous measurement of soil moisture under field conditions. Plant Physiology. 1948;**23**(4):532-543. DOI: 10.1104/pp.23.4.532

[59] Huang SC, Lin YZ. A low-cost constant-moisture automatic irrigation system using dynamic irrigation interval adjustment. Applied Science. 2020;**10**:6352. DOI: 10.3390/app10186352

[60] Sentelhas PC, Monteiro JE, Gillespie TJ. Electronic leaf wetness duration sensor: Why it should be painted. International Journal of Biometeorology. 2004;**48**(4):202-205. DOI: 10.1007/s00484-004-0200-z

[61] Patle KS, Saini R, Kumar A, Surya SG, Palaparthy VS, Salama KN. IoT enabled, leaf wetness sensor on the flexible substrates for in-situ plant disease management. IEEE Sensors Journal. 2021:1-1. DOI: 10.1109/jsen.2021.3089722

Chapter 4

Oxidative Stability and Sensory Properties of Pecan Nuts

Adriana María Descalzo, Sergio Aníbal Rizzo, Carolina Daiana Pérez, Andrea Biolatto, Enrique Alberto Frusso, Gabriela María Grigioni and Luciana Rossetti

Abstract

Pecans are the nut with the higher oil content. In addition, they present a large number of polyunsaturated fatty acids, which are susceptible to oxidation. Oxidative damage in pecans is traduced in lower quality aspects, appearance of rancidity and acidity, loss of sweetness and firmness, darker kernels, and darker shells. The use of different strategies for the conservation of entire and shelled nuts is discussed in terms of oxidation and the consequences on nuts quality.

Keywords: pecan, *Carya illinoinensis*, oxidation, volatiles, sensory, postharvest, tocopherols, antioxidants

1. Introduction

Pecans nuts are the seeds of *Carya illinoinensis* (Wangenh.) K. Koch. They are an important source of vitamins and minerals, such as vitamin E, folic acid, calcium, magnesium, phosphorus, potassium, several B vitamins, and zinc. They are also rich in fatty acids, having around 58.1 to 68.18 g oil/100 g of nut. Particularly, the unsaturated/saturated fatty acids ratio is around 13.54, with 93% of unsaturated fatty acids in the oil [1].

They also contain β-sitosterol and squalene ranging from 88.74 to 220.42 mg 100 g^{-1} and 30.98 to 115.59 mg 100 g^{-1}, respectively [2]. This is a disadvantage for the conservation of the nuts and oils because unsaturated fatty acids are prone to oxidative damage. But on the other side, they contain a high level of γ-tocopherol (the main form of vitamin E in pecans) and polyphenols, especially epigallocatechin-3-gallate (EGCG), recognized as a health promoter compound [3] and ellagic acid associated with hepatoprotective activity [4]. Recently, the determination of the proximal composition in 11 cultivars of Brazilian pecans [2] showed total phenolic content variations ranging from 19.88 mg GAE g−1 ("Desirable" pecans) to 45.25 mg GAE g−1 ("Imperial").

 IntechOpen

Pecans are also a source of protein between 6.88 and 9.26 g .100 g^{-1} nut with low carbohydrate content (between 4.92 and 17.33 g 100 g^{-1} nut), and dietary fiber (5.55 to 15.94 g 100 g^{-1} nut).

In terms of dietary issues, pecans have the lowest net carbohydrates content of any nuts. In fact, one ounce of pecans contains just 1.1 grams of net carbohydrates. This means its consumption is possible, while following a low-carb diet.

2. Lipid composition in terms of oxidative stability

Pecan nuts are rich in fats, with a net yield of around 58.1–66.18 g oil/100 g of nut mass. Six classes of lipids were separated and identified as complex lipids, monoglycerides, α-β-diglycerides, α-α'-diglycerides, sterols, and triglycerides. Triglycerides were predominant with a mean concentration for the six cultivars equal to 71.25 g/100g of nutmeat [5]. The total content of unsaturated fatty acids in the oil is as high as 93%. The unsaturated fatty acids are a group of lipids containing one or more double bonds in the structure. Monounsaturated fatty acids (MUFAs) contain one double bond in the fatty acid chain. If two or more double bonds are present, they are referred to as polyunsaturated fatty acids (PUFAs).

The number of double bonds has an impact in terms of cell membrane mobility. The higher PUFAs, the higher lipid mobility within membranes. Saturated fatty acids (SFAs) do not contain double bonds in their structure and contribute to a more rigid cellular structure. The low ratio of saturated to unsaturated fatty acids in the cell membranes increases membrane fluidity and permeability [6].

Omega-6 linoleic acid and omega-3 alpha-linolenic are essential fatty acids that cannot be synthesized endogenously by most of the animals; therefore, their constant dietary intake is crucial [7].

The chromatographic profile of Stuart pecans [6] fatty acids is shown in **Figure 1**.

The largest peak corresponds to oleic acid (C18: 1 n-9), followed by linoleic acid (C18: 2 n-6). This picture is typical of the fatty acids profile of pecan products (Picture from the Food Science laboratory INTA). Indeed, pecan varieties show a

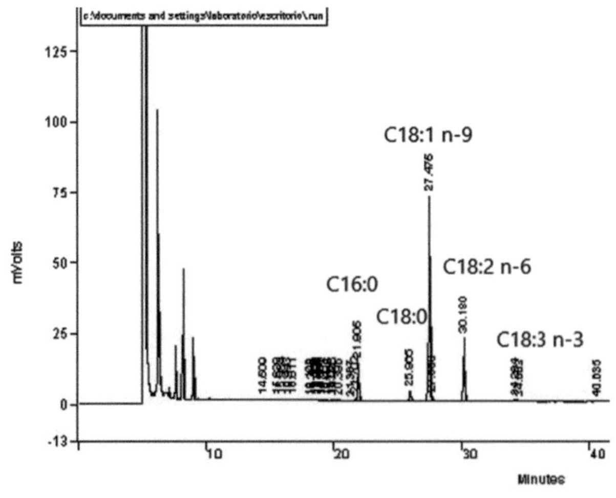

Figure 1.
Gas chromatography (GC-FID) typical profile of pecan nuts.

DOI: http://dx.doi.org/10.5772/intechopen.106175

		Main fatty acids								
Country	Variety	C16:0	C18:0	C18:1n9	C18:2n6	C18:3n3	ΣSFA	ΣUFA	ΣPUFA	PI
BR	Barton	6.53 ± 0.37 a	2.46 ± 0.31 abcd	66.92 ± 0.27 ab	22.50 ± 0.22 abc	0.93 ± 0.01 a	9.0	90.4	23.4	41.1
BR	Chickasaw	5.64 ± 0.18 a	3.07 ± 0.25 abcd	72.86 ± 0.10 a	16.91 ± 0.14 c	0.86 ± 0.01 a	8.7	90.6	17.8	36.8
BR	Desirable	5.68 ± 0.21 a	3.64 ± 0.27 abc	69.79 ± 0.25 ab	19.38 ± 0.13 c	0.87 ± 0.01 a	9.3	90.0	20.3	38.6
BR	Imperial	6.10 ± 0.06 a	3.20 ± 0.15 abcd	68.69 ± 0.11 ab	20.58 ± 0.04 abc	0.82 ± 0.01 a	9.3	90.1	21.4	39.4
BR	Importada	6.46 ± 0.15 a	2.66 ± 0.15 abcd	68.57 ± 0.97 ab	19.81 ± 0.26 bc	1.06 ± 0.02 a	9.1	89.4	20.9	39.1
BR	Jackson	6.48 ± 0.07 a	4.01 ± 0.07 a	68.12 ± 0.14 abc	19.99 ± 0.20 bc	0.76 ± 0.01 a	10.5	88.9	20.8	38.5
BR	Mahan	5.66 ± 0.13 a	3.49 ± 0.29 abcd	72.99 ± 0.05 a	16.27 ± 0.12 c	1.02 ± 0.01 a	9.2	90.3	17.3	36.6
BR	Melhorada	6.40 ± 0.21 a	3.82 ± 0.20 ab	68.40 ± 0.29 abc	19.77 ± 0.07 bc	0.98 ± 0.03 a	10.2	89.2	20.8	38.8
BR	Moneymaker	5.80 ± 0.12 a	3.34 ± 0.23 abcd	69.50 ± 0.13 ab	19.86 ± 0.05 bc	0.86 ± 0.01 a	9.1	90.2	20.7	39.0
BR	Stuart	5.68 ± 0.08 b	2.414 abcd	69.42 ± 0.38 bc	18.54 ± 0.40 e	1.02 ± 0.05 ab	10.4	89.0	19.6	37.9
BR	Success	6.40 ± 0.36 a	4.44 ± 0.23 abcd	65.53 ± 0.43 abc	22.06 ± 0.12 abc	0.90 ± 0.01 de	10.8	88.5	23.0	40.2
US	Western	8.18 ± 1.66 a	2.0 ± 0.66 bcd	54.33 ± 5.65 c	34.08 ± 5.94 a	1.4 ± 0.77 a	10.2	89.8	35.5	50.5
US	Barton	7.53 ± 2.63 a	2.1 ± 0.49 bcd	66.3 ± 11.83 abc	22.45 ± 9.62 abc	1.43 ± 0.36 a	9.6	90.2	23.9	41.9
US	Success	6.85 ± 0.69 a	1.93 ± 0.42 abcd	63.15 ± 7.06 abc	29.18 ± 7.82 abc	1.45 ± 0.41 a	8.8	93.8	30.6	47.9
US	Stuart	7.23 ± 1.26 a	1.85 ± 0.38 abcd	61.28 ± 5.9 abc	25.5 ± 5.51 abc	1.53 ± 0.53 a	9.1	88.3	27.0	43.9
US	Schley	7.9 ± 1.01 a	1.8 ± 0.42 cd	59.4 ± 3.00 abc	29.63 ± 1.80 abc	1.38 ± 0.57 a	9.7	90.4	31.0	47.2
US	Texas Prolific	7.1 ± 1.24 a	1.95 ± 0.79 cd	57.1 ± 5.12 bc	33.35 ± 5.18 ab	1.325 ± 0.62 a	9.1	91.8	34.7	50.3
US	Hayes	7.67 ± 1.74 a	1.65 ± 0.52 d	59.75 ± 5.74 abc	29.5 ± 6.69 abc	1.3 ± 0.7 a	9.3	90.5	30.8	47.0
AR	Stuart	6.84 ± 0.54 a	2.52 ± 0.12 abcd	60.55 ± 1.06 abc	26.63 ± 0.78 abc	1.09 ± 0.08 a	9.4	88.3	27.7	43.9

SFA (saturated fatty acids); UFA (total Unsaturated fatty acids); PUFA (Polyunsaturated fatty acids); Peroxidizability Index (PI) = [(%Monoenoic × 0.025) + (%Dienoic × 1) + (%Trienoic × 2)], as stated previousl [9]. Numbers in yellow indicate the highest levels and in gray the lowest values.
Different letters within column indicate significant differences ($p < 0.05$).

Table 1.
Main fatty acids in pecan kernels from Brazil, US and Argentine varieties (g/100 g fat).

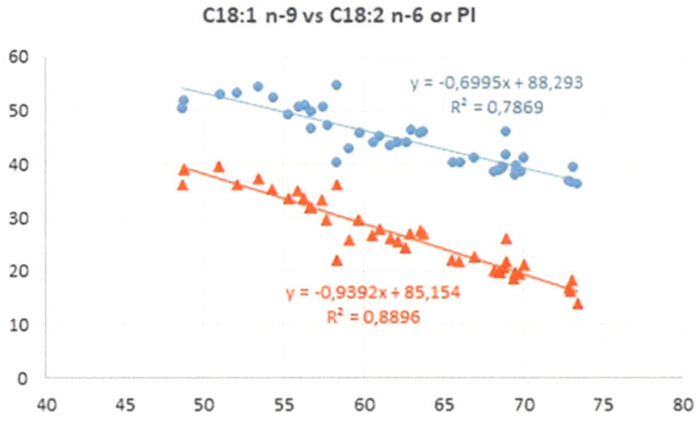

Figure 2.
Relationship between oleic acid, linoleic acid and Peroxidizability Index. Dots in blue indicate C18: 1 n-9 vs. PI; triangles in red indicate C18: 1 n-9 vs. C18: 2 n-6.

conservative fatty acids profile with slightly different shapes attributable mainly to the genetic, environmental, and plant phenological status [8]. In **Table 1**, the means of major fatty acids are shown in different cultivars from different regions (US, Brazil, and Argentina).

As lipid composition is critical for oxidative stability, different fatty acids and their indexes are presented in **Table 1**. Oleic acid (C18: 3, n-9) is the main constituent of pecan fatty acids. It has one double bond at the carbon in position nine. Therefore, this FA is relatively stable in terms of oxidation. Its concentration in pecan kernels is inversely proportional to the linoleic acid (C18: 2, n-6) concentration (**Figure 2**) and consequently to the peroxidizability index (PI).

The peroxidizability index may indicate the susceptibility of pecans to oxidation. At higher levels of oleic acid, the lower PI in the samples. Mahan showed the lowest total PUFA level and consequently the lowest PI. The variation in the lipid content affected the pecan stability, as oxidative damage induces quality loss and correlates positively with rancidity [8].

Taking together all the data in **Table 1**, it is noticeable that the samples with higher oleic acid (C18: 1 n-9) presented lower levels of linolenic acid (C18: 2 n-6), that is, Mahan vs. Western with a negative and significant linear correlation (R2 = 0.89), as illustrated in **Figure 2**.

By virtue of the number of peroxidizable double bonds, samples with higher oleic acid are more stable and show less PI. Apparently, during the development, oleic acid is transformed into linoleic acid by the action of desaturates. As the proportion of oleic acid increased or decreased, it was in tandem with an opposite change in linoleic acid. The ratio oleic/linoleic coincided with the ratio oleic/PI, indicating the susceptibility of the cultivars [8]. Therefore, it can be considered for breeding programs or further postharvest handling in terms of oxidative stability.

Desirable pecans grown at two different locations (BW and CW) showed differences in their oleic-linoleic composition. The fatty acid profile may depend on environmental conditions, cultivar, maturity, and horticultural practices. Thus, the selection of appropriate cultivation conditions is an important factor to consider when selecting pecan cultivars [10].

For example, increasing the nitrogen fertilization rate increases the protein content and oleic acid levels in kernels. At the same time, it showed a tendency to lower tocopherol contents [11].

	Pecan oil Variety									
Fatty acid	**Shoshoni**	**Desirable**	**Kernodle**	**Success**	**Mahan**	**Starking**	**Stuart**	**Cheyenne**	**INTA Delta I**
Palmitic acid (C16:0)	5.79 ± 0.07 d	5.40 ± 0.02 f	5.65 ± 0.10 de	5.52 ± 0.12 ef	6.11 ± 0.10 c	6.68 ± 0.01 a	6.33 ± 0.04 bc	6.42 ± 0.03 b	6.69 ± 0.13 a
Palmitoleic acid (C16:1)	0.07 ± 0.01 cd	0.08 ± 0.01 bc	0.07 ± 0.00 d	0.10 ± 0.01 a	0.07 ± 0.00 cd	0.08 ± 0.00 bc	0.08 ± 0.00 bcd	0.08 ± 0.00 bc	0.09 ± 0.00 b
Margaric acid (C17:0)	0.05 ± 0.00 a	0.06 ± 0.00 a	0.06 ± 0.01 a	0.06 ± 0.00 a	0.06 ± 0.00 a	0.06 ± 0.01 a	0.06 ± 0.00 a	0.06 ± 0.00 a	0.07 ± 0.01 a
Heptadecanoic acid (C17:1)	0.05 ± 0.00 abc	0.06 ± 0.00 ab	0.06 ± 0.01 ab	0.06 ± 0.00 a	0.05 ± 0.01 bc	0.04 ± 0.00 c	0.05 ± 0.00 abc	0.05 ± 0.00 bc	0.06 ± 0.01 abc
Esteáric acid (C18:0)	2.47 ± 0.02 e	2.72 ± 0.02 cd	1.96 ± 0.03 h	2.82 ± 0.04 b	3.02 ± 0.02 a	2.65 ± 0.01 d	2.22 ± 0.04 g	2.34 ± 0.02 f	2.78 ± 0.01 bc
Oleic acid (C18:1 trans)	0.03 ± 0.00 a	0.01 ± 0.01 a	0.02 ± 0.01 a	0.02 ± 0.00 a	0.03 ± 0.01 a	0.02 ± 0.01 a	0.02 ± 0.01 a	0.02 ± 0.01 a	0.02 ± 0.01 a
Oleic acid (C18:1 cis n-9)	62.64 ± 0.10 c	67.71 ± 0.04 a	62.26 ± 0.41 c	65.01 ± 0.11 b	59.63 ± 0.02 e	49.44 ± 0.01 h	60.99 ± 0.06 d	52.52 ± 0.04 g	54.54 ± 0.09 f
Linoleic acid (C18:2 n-6)	27.27 ± 0.07 f	22.56 ± 0.02 h	28.40 ± 0.32 e	25.07 ± 0.06 g	29.37 ± 0.05 d	38.94 ± 0.02 a	28.61 ± 0.03 e	36.77 ± 0.01 b	33.90 ± 0.08 c
Linolenic acid (C18:3 n-3)	1.20 ± 0.00 e	0.99 ± 0.01 g	1.12 ± 0.01 f	0.91 ± 0.00 h	1.29 ± 0.02 d	1.73 ± 0.01 a	1.22 ± 0.01 e	1.33 ± 0.00 c	1.47 ± 0.00 b
Araquidonic acid (C20:0)	0.13 ± 0.01 c	0.14 ± 0.01 abc	0.10 ± 0.00 d	0.15 ± 0.00 a	0.14 ± 0.00 a	0.14 ± 0.00 abc	0.12 ± 0.01 c	0.13 ± 0.00 bc	0.14 ± 0.00 ab
Erudic acid (C20:1)	0.27 ± 0.01 a	0.26 ± 0.01 a	0.27 ± 0.01 a	0.26 ± 0.00 a	0.20 ± 0.00 bc	0.18 ± 0.01 c	0.27 ± 0.01 a	0.28 ± 0.01 a	0.21 ± 0.00 b
Behenic acid (C22:0)	0.03 ± 0.00 ab	0.02 ± 0.00 b	0.03 ± 0.00 ab	0.02 ± 0.00 b	0.03 ± 0.01 ab	0.03 ± 0.00 a	0.03 ± 0.00 ab	0.03 ± 0.00 ab	0.03 ± 0.00 ab

Different letters within column indicate significant differences ($p < 0.05$).

Table 2.
Fatty acids profile in pecan oil from Argentine cultivars (g/100 g pressed oil).

As expected, the fatty acids profile in pecan oil resembles those from kernels, as fat is the major constituent of pecan nuts (**Table 2**).

As shown in **Table 2**, results in extracted pecan oil, the desirable variety had the higher oleic/linoleic relationship, whereas starking showed the lowest, as demonstrated for the pecan oil extracts [12].

Oil is highly susceptible to oxidation, because it does not have the protective structure of the nut proteins and polyphenols, with a shell containing antioxidants. The presence of a large number of phenolic compounds in pecan nut shells may explain the protective activity.

3. Antioxidants in pecan shell, oil, and nuts

3.1 Shells

Recently, the presence of 29 and 27 phenolic compounds in nut shells in aqueous and hydro-alcoholic phases has been reported, respectively, from samples submitted to ultrasonic-assisted extraction. The major compounds resulted in catechin (>260 mg/g dry sample); gallic acid (>120 mg/g dry sample); epicatechin (>24 mg/g dry sample); myricetin (>12 mg/g dry sample); ellagic acid (>11 mg/g dry sample); and vanillin (> 6 mg/g dry sample). Other minor compounds are chlorogenic acid, vanillic acid, siryngic acid, epicatechingallate, fustin, P-coumaric acid, taxifolin, ferulic acid, rosmarinic acid, quercetin, salicylic acid, myricetin, eriodictyol, naringenin, and galangin [13].

3.2 Nuts and oil

The main antioxidant in pecan oil is γ-tocopherol 24.97 ± 0.90 (mg/100 g), followed by α-tocopherol 1.65 ± 0.02 (mg/100 g). The presence of γ-tocopherol and a high concentration of oleic acid are associated with the high oxidative stability of pecan nut oil [14]. In previous work, we described that in Stuart pecans, tocopherols ranged from 5.8 to 142 mg kg^{-1} nut of α- and γ- tocopherol, respectively. Minor quantities of β- + δ- tocopherols were also detected (less than 0.5 $mg.kg^{-1)}$ [6].

Phytosterols are bioactive compounds that act on oxidative stability in plant cell membranes. They have in their chemical structure, a steroid nucleus with hydroxyl groups (3-β-hydroxyl group), which could be related to a mild antioxidant activity exerted in the lipid phase of biological membranes [15]. β-sitosterol is the main component of pecan sterols (approximately 75%), and its concentration is variable and depends on their ripening state. In Tunisian pecans, the changes in 4-desmethylsterol, 4-monomethylsterol, 4, 4-dimethylsterol, and phytostanol composition were quantitatively and qualitatively investigated during the ripening of three varieties (Mahan, Moore, and Burkett). Fifteen phytosterols and one phytostanol were quantified. The greatest amount of phytosterols (2852.5 mg/100 g of oil) was detected in the Mahan variety, 20 weeks after the flowering date (WAFD). Moore had the highest level of phytostanols (7.3 mg/100 g of oil) at 20 WAFD. Phytosterol and phytostanol contents showed a steep decrease during pecan nut development. Results from the quantitative characterization of pecan nut oils revealed that β-sitosterol, Δ5-avenasterol, and campesterol were the most abundant phytosterol compounds at all ripening stages [16]. Also, they protected oil and fruits from oxidative processing. It has been shown that in fried pecans, tocopherols and phytosterols were the main functional components in the oil-soluble part, to avoid oxidation [17].

The values for total carotenoids in cultivars varied from 0.897 to 1.403 μg/g of oil without any significant difference among cultivars for total carotenoid content [18].

Some antioxidants reported in the literature are extracted in **Table 3**.

C. illinoensis (variety)	Total Phenolic Activity (mg CAE/g)	γ-tocopherol (μg/g)	Phytosterols (PE) (μg/g)	Reference
Kanza	106 ± 2.3	105 ± 1		Oil [10]
Nacono	76 ± 2.2	135 ± 4		
Kiowa	76 ± 2.5	72 ± 6		
Pawnee	72 ± 0.9	100 ± 1		
Shawnee	71 ± 1.9	102 ± 2		
Desirable CW	70 ± 2.0	84 ± 1		
Desirable BW	62 ± 2.3	126 ± 5		
Stuart		139 ± 3		kernel
Sioux		311 ± 3		[19]
Pawnee		86.8 ± 3		
Cheyenne	21.6			Kernel
Choctaw	23			[20]
Desirable	19.5			
Pawnee	20			
Stuart	22.7			
Summer	19.3			
Western	21.9			
Wichita	24.4			
MV (mixed varieties 2009)	130.30 ± 3.45 (shell extract)	381 ± 4	2200	Oil and shell extract [21]
MV (mixed varieties 2010)	145.41 ± 8.16 (shell extract)	238 ± 5	2100	
Barton (2009)	94.04 ± 2.66 (shell extract)	334 ± 41	1900	
Barton (2010)	181.49 ± 6.97 (shell extract)	260 ± 10	1900	
Stuart		393		Oil [8]
Sioux		291		
Choctaw		221		
Barton		410		
Shawnee		369		
Desirable		313		
C. cathayensis	19.4	72.1	β-sitosterol: 728 stigmasterol:144 fucosterol:101 campsterol: 21.3	Kernel [17]

Table 3.
Main antioxidants in pecans: γ-tocopherol, phytosterols and total phenols.

4. Oxidative stability

The conservation of pecan nuts is important in terms of their quality because the sensory defects are undesired for commercial purposes and further consumption.

Many authors described the oxidative stability of nuts with accelerated assays. Indeed, the results may be useful to understand the relationship between biochemical and sensory properties, but they do not predict properly the mechanisms under real storage conditions, because higher temperatures may accelerate lipid oxidation/degradation and Maillard reactions with the concomitant formation of Strecker products that are not expected or are less produced under low-temperature storage [22].

Sensory defects can be predicted in pecans by multiple biochemical indicators. Rancid flavor, moisture, and conjugated dienes were representative of quality deterioration, whereas secondary oxidation products related to higher thiobarbituric acid reactive substances (TBARS) [23]. Other authors indirectly associated the odor of treated pecans with oxidative deterioration and rancid flavor using an e-nose with a sensor array [24]. Recently, a report indicated that storage temperatures below 10°C were more effective to preserve Barton pecans than low oxygen atmospheres with 20, 3 and 1.5 Kpa O2 [25]. In addition, the breakdown of flavonoids and reaction products from Maillard browning could be responsible for the formation of the reddish-brown

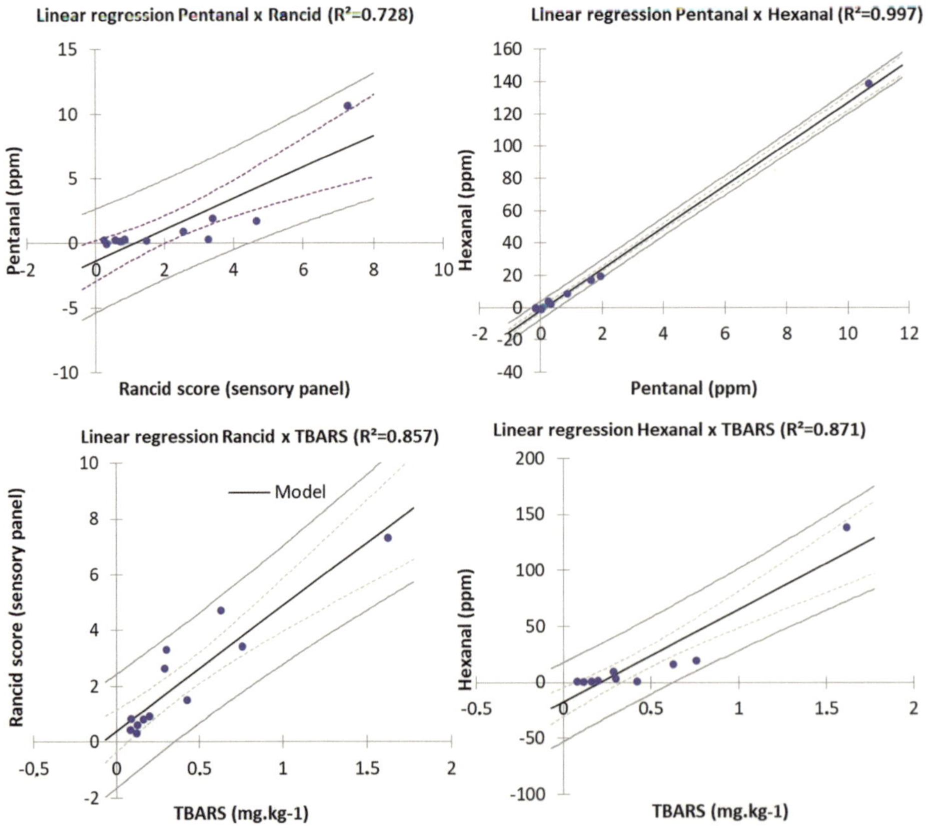

Figure 3.
Linear regression between hexanal, pentanal and TBARS as predictors for rancid flavor in Stuart pecans.

color observed in degraded nutmeats. Browning can be predicted using a mathematical model with a first-order decay equation [26].

Moreover, the conservation of pecan nuts for 10 months postharvest is interesting because it allows producers to commercialize their product until the next harvest period. They may be stored in clean burlap bags, at a maximal atmosphere humidity of 70%, and in a dark, clean, and ventilated chamber.

In previous work, we have described that the TBARS value (thiobarbituric acid reactive substances), hexanal, and pentanal were the main predictors of pecan oxidative stability [6]. The degradation of linolenic acid produced hexanal and pentanal as major volatile aldehydes [23, 27] as a result of the cleavage and oxidation of the double bonds. Hexanal and pentanal correlated positively between them, indicating a similar occurrence. Both volatiles were produced in the samples but at different concentrations.

Interestingly, both volatiles showed a positive and significant correlation with TBARS and rancid score, as determined by a sensory panel [6]. Some correlations are shown in **Figure 3**.

This feature may indicate that a TBARS value of 1 corresponds to a rancid score of 5, using a 10-cm nonstructured scale where the lower extreme meant "extremely weak" and the upper extreme meant "extremely strong." Therefore, a TBARS value equal to 1 (mg.kg-1) can be signaled as the threshold value to perceive pecan kernels as rancid.

Oxidation progress within the storage of pecans is independent of the concentration of tocopherols content in the samples. Both α- and γ-tocopherols are preserved during unshelled pecans storage at either refrigerated or room temperature, whereas oxidation triggering measured by TBARS started at seven months postharvest, regardless of the storage temperature [6].

5. Conservation of pecans and sensory attributes

Many authors describe different strategies for the conservation of nuts. These strategies comprise shelled nuts with different packaging and atmosphere environments, as well as in shell conservation, mainly for in-bulk storage in pecan facilities under different temperatures and oxygen conditions. The compromise between energy cost and pecan quality maintenance will depend on producers' and retailers' possibilities.

5.1 Shelled nuts under vacuum storage at room temperature

In a previous work, shelled nuts were submitted to room temperature storage, either within nylon-polyethylene bags, or in polypropylene containers. They were conserved for 150 days. Peroxides were raised in both treatments as well along with the time of storage. They described a significant linear reduction ($p < 0.0001$) in all the sensory characteristics (visual color, typical flavor, odor, and texture) during the storage period. The acceptability scores were similar for both types of packaging, indicating that vacuum treatment did not contribute to prolonging the pecan's shelf-life [28]. The shelf life for both treatments was determined in 120 days of storage. After that time, sensory scores dropped below six (scale 1 to 10).

5.2 Raw and roasted shelled pecans stored at room temperature

Roasting promoted the oxidation of the lipids, with higher TBARS and peroxides levels compared with raw nuts. However, the nonsignificant differences in rancid aroma and rancid flavor between raw and roasted pecans implied that the sensory response was not sensitive enough to differentiate any increase in oxidative products at an early stage of storage. In this study, panelists did not differentiate the lightness of the raw and roasted products, whereas darkening was indicated in roasted pecans using instrumental measurements.

Storage affected the crunchiness of raw and roasted pecans, with those stored at 65% relative humidity (RH), having lower scores than those stored at 55% RH. During storage, rancid aroma and flavor developed in both raw and roasted pecans with higher records for roasted ($p < 0.05$). Varying the relative humidity, 55% and 65% RH, during storage, did not affect flavor or aroma scores significantly for either raw or roasted pecans [23].

5.3 Shelled coated pecans

Shelled "Desirable" pecans were coated with different mixtures of (MC = methylcellulose, CMC = carboxymethyl cellulose, HPC = hydroxypropyl cellulose, PG = propylene glycol, BHA = butylated hydroxyanisole, BHT = butylated hydroxytoluenecaboxy methyl cellulose). All types of coating preserved pecan kernels from oxidative damage compared with uncoated pecans. Coated kernels initially had slightly higher off-flavor, perhaps due to the coating itself, but had less off-flavor and better overall flavor after nine months of storage. Also, the hexanal levels after five months of storage were twofold less in coated than in uncoated kernels. For that reason, the coating could reduce lipid oxidation (i.e., rancidity) and preserve the color during marketing at ambient temperature by limiting oxygen contact with the kernel lipids. This strategy would reduce costs to the pecan industry and improve quality for the consumers [29].

5.4 Shelled pecans stored at different temperatures and different packaging conditions

"Stuart" shelled pecans were stored for 12 months at 20°C, 2°C, and – 15°C packaged in cellophane bags or bi-laminated polythene-aluminum bags flushed with nitrogen. The rise in oxidation (TBARS) between the fourth and tenth month was 10-fold, 4-fold; 4.6-fold, and 4.6-fold for 20°C; 2°C, and – 15°C, respectively. Under nitrogen, oxidation showed no differences between temperatures. TBARS increase was 5.0-fold, 4.6-fold, and 3.7-fold for 20°C; 2°C, and – 15°C, respectively (**Figure 4**). Differences in lipid oxidation were also noticeable in color (**Figure 5**). Nuts stored in cellophane bags were less conserved than under nitrogen. This assay demonstrated that the use of bilaminated bags and nitrogen atmosphere conserved the quality of the nuts, without the necessity of using low-temperature storage. Without nitrogen, refrigeration or freezing were the suitable options. Unshelled pecans showed higher oxidation and darker color after 10 months of storage (Experimental data from INTA).

5.5 Shelled and unshelled pecans stored at 20°C and –15°C temperatures under modified atmosphere conditions

Dried shelled and unshelled "Barton" pecans were separated into two parts, one submitted to 1-MCP (SmartFresh®, 0.14% of active ingredient as maturation

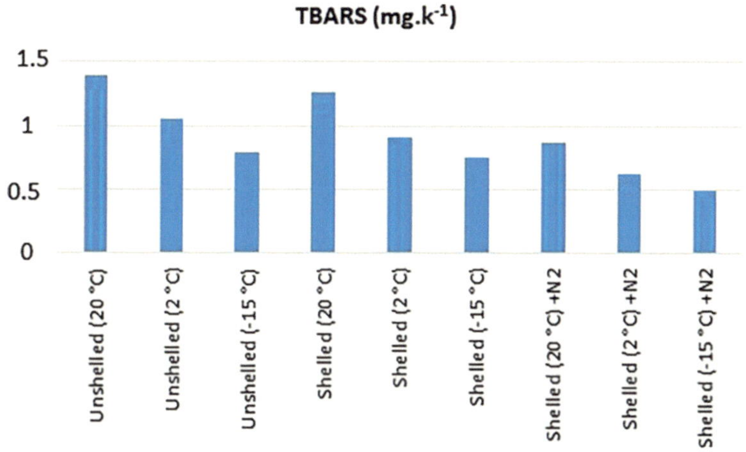

Figure 4.
TBARS levels of shelled pecans stored under different packaging and temperature conditions. TBARS assay at the laboratory of Food Science and Technology, INTA.

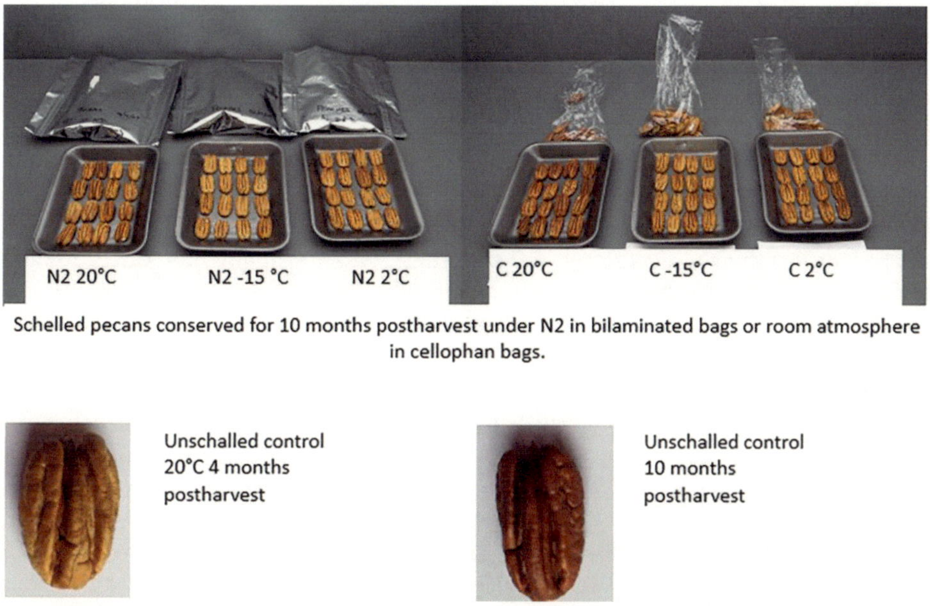

Figure 5.
Aspect of pecans stored in different bags, with or without nitrogen and different temperatures. Color assay at the laboratory of Food Science and Technology, INTA.

inhibitor) at a concentration of 1.0 μL L^{-1}. Four groups (shelled and unshelled, with or without MCP-1 treatment) were stored under ambient temperature (17 ± 5.1°C) and relative air humidity (78.3 ± 11.2%). Each treatment was identified and placed separately into raffia bags until evaluation.

Since pecans respire and continue to have enzymatic activity throughout storage unless deactivated using heat or irradiation, it is essential to maintain the packaging environment that can slow down the rate of deterioration.

Results showed that 1-MCP application inhibited excessive acidity and lipid peroxidation in both shelled and unshelled nuts. Luminosity was better conserved in unshelled pecans. Also, some unwanted aldehydes were produced at lower levels, indicating some protective action of the shell. Nonetheless, 1-MCP treatment in this condition reduced the abundance of these volatile compounds.

The conservation of the shells prevented oxidative damage in the nuts. In summary, unshelled pecans stored under ambient conditions and with a 1-MCP application showed the best quality [30].

5.6 Unshelled pecans stored at room temperature and 2°C

Unshelled "Stuart" pecans were dried to a moisture content of 3–4% and placed in individual clean mesh bags, conserved either at 2 ± 1°C or at 20 ± 1°C at 65% of relative humidity in the dark for 10 months. These conditions resemble postharvest practices in pecan facilities.

Refrigeration did not avoid the trigger of oxidation, measured by TBARS, hexanal, and pentanal. But it restricted the final levels of oxidation compounds. Refrigerated storage of in-shell pecans resulted in differences detected at both biochemical and sensory scores, with significantly higher signs of oxidative deterioration at 20°C compared with storage temperature at 2°C.

Oxidative damage showed an exponential evolution that triggered from day 210 at 20°C and 30 days later at 2°C. A similar behavior was observed for the presence of rancid taste and typical flavor loss during postharvest storage.

The conservation of unshelled nuts at 2°C prevented the excessive formation of these compounds, compared with the conservation at 20°C, but it did not prevent the initiation of the oxidative process. Moreover, the results of sensory trials showed that pecans lost their typical flavor and sweetness, whereas augmented the bitter and rancid taste along with the storage with enhanced deterioration at 20°C compared with 2°C [6].

5.7 In shell pecans are stored at room temperature, refrigerated storage, and different oxygen partial pressure (pO2)

"Barton" unshelled pecans were stored in containers and placed in three rooms at 20, 10, and 1.5°C and under three different pO2, 20 kPa (ambient condition), 3 kPa, and 1 kPa. The ambient temperature (20°C) allows for avoiding refrigeration. Refrigerated storage (8–10°C) is currently adopted by companies. The lowest temperature used as a control was 1.5°C. After 12-month storage, pecans kept at room temperature (20°C) showed increased acidity and color change.

Luminosity decreased at 20°C with respect to other refrigeration conditions. However, samples kept at 20°C and with lower pO2 (3 and 1 kPa) maintained higher luminosity and less oxidation indicators throughout storage.

Therefore, adopting lower temperatures (1.5 and 10°C) resulted more effective at maintaining quality regardless of the atmosphere condition, without any significant differences in the luminosity and the presence of volatile aldehydes and acids production.

The use of low pO2 in storage facilities has shown positive results, especially at higher temperatures (20°C). There was little difference in quality between 1 and 3 kPa; thus, pO2 near to 3 kPa can be recommended, especially when there are required lower energy costs with refrigeration [24].

6. Conclusions

Pecans are nuts rich in oil and, especially in unsaturated fatty acids making them prone to oxidation. Oxidative damage leads to a decrease in overall quality, appearance of rancidity, loss of sweetness, typical flavor, darkening, and texture defects. Unshelled pecans stored at room temperature begin oxidizing between 6 and 8 months postharvest. This can be avoided or diminished using low-temperature of storage until reaching the next harvest period. Other strategy for retail is the storage of shelled pecans, but they are more susceptible to oxidation than shelled nuts because the shell contains many phenolic antioxidants that can form a protective barrier against oxidation. For that reason, shelled nuts may be submitted to coating, packaging, and temperature or the use of a modified atmosphere to preserve the quality of kernels.

The use of nitrogen or atmosphere modifiers allowed to store of unshelled pecans at room temperature. Research on packaging and atmosphere may be also a key factor in the storage and distribution of bulk nuts in 25 kg bags.

Acknowledgements

The work is part of the INTA National Program of Agroindustry, added value, and bioenergy, supported by the projects PEI517: Food Quality. Sensory and analytical attributes and the PD152: Functional Foods.

Conflict of interest

The authors declare no conflict of interest.

Author details

Adriana María Descalzo[1,2]*, Sergio Aníbal Rizzo[1], Carolina Daiana Pérez[1,2], Andrea Biolatto[1], Enrique Alberto Frusso[1], Gabriela María Grigioni[1,2] and Luciana Rossetti[1]

1 National Institute of Agricultural Technology (INTA), Argentina

2 National Scientific and Technical Research Council (CONICET), Argentina

*Address all correspondence to: descalzo.adriana@inta.gob.ar

References

[1] Ryan E, Galvin K, O'Connor TP, Maguire AR, O'Brien NM. Fatty acid profile, tocopherol, squalene and phytosterol content of brazil, pecan, pine, pistachio and cashew nuts. International Journal of Food Science and Nutrition. 2006;**57**:219-228

[2] Reis Ribeiro S, Klein B, Machado Ribeiro Q, dos Duarte Santos I, Gomes Genro AL, de Freitas Ferreira D, et al. Chemical composition and oxidative stability of eleven pecan cultivars produced in southern Brazil. Food Research International. 2020;**136**:109596

[3] Venkatachalam M, Sathe SK. Chemical composition of selected edible nut seeds. Journal of Agricultural and Food Chemistry. 2006;**54**:4705-4714

[4] Watson RR. Dietary Interventions in Liver Disease: Foods, Nutrients and Dietary Supplements. 1st ed. San Diego, CA: Elsevier; 2018

[5] Senter SD, Hdrvat RJ. Lipids of Pecan Nutmeats. Journal of Food Science. 1976;**41**:1201-1203

[6] Descalzo AM, Biolatto A, Rizzo SA, Pérez CD, Frusso EA, Carduza F, et al. Oxidative stability parameters and sensory properties of in-shell "Stuart" pecans [Carya illinoinensis (Wangenh.) K.Koch] stored at different temperatures under non-accelerated conditions. Postharvest Biology and Technology. 2021;**179**:111591

[7] El-Badry AM, Moritz W, Contaldo C, Tian Y, Graf R, Clavien PA. Prevention of reperfusion injury and microcirculatory failure in macrosteatotic mouse liver by omega-3 fatty acids. Hepatology. 2007;**45**:855-863

[8] Rudolph CJ, Odell GV, Hinrichs HA, Hopfer DA, Kays SJ. Genetic, environmental, and maturity effects on pecan kernel lipid, fatty acid, tocopherol, and protein composition. Journal of Food Quality. 1992;**15**:263-278

[9] Witting LA, Horwitt MK. Effect of degree of fatty acid unsaturation in tocopherol deficiency-induced creatinuria. The Journal of Nutrition. 1964;**82**:19-33

[10] Villarreal-Lozoya JE, Lombardini L, Cisneros-Zevallos L. Phytochemical constituents and antioxidant capacity of different pecan [Carya illinoinensis (Wangenh.) K. Koch] cultivars. Food Chemistry. 2007;**102**:1241-1249

[11] Frusso EA. Influencia del nitrógeno, fósforo y cinc sobre la composición química y rendimiento de la nuez pecán y su relación con la variabilidad de nutrientes en hoja. Argentina: Buenos Aires University; 2013

[12] Panozzo A, Consiglio Diez L, Urbani V, Zapata LM. Composición de ácidos grasos de aceites de distintas variedades de nuez pecan [Carya illinoinensis (Wangenh.) K. Koch] cultivadas en la provincia de Entre Ríos. 2019

[13] Hilbig J, Alves VR, Müller CMO, Micke GA, Vitali L, Pedrosa RC, et al. Ultrasonic-assisted extraction combined with sample preparation and analysis using LC-ESI-MS/MS allowed the identification of 24 new phenolic compounds in pecan nut shell [Carya illinoinensis (Wangenh) C. Koch] extracts. Food Research International. 2018;**106**:549-557

[14] Polmann G, Badia V, Frena M, Teixeira GL, Rigo E, Block JM, et al. Enzyme-assisted aqueous extraction

combined with experimental designs allow the obtaining of a high-quality and yield pecan nut oil. LWT. 2019;**113**:108283

[15] Yoshida Y, Niki E. Antioxidant effects of phytosterols and its components. Journal of Nutritional Science and Vitaminology (Tokyo). 2003;**49**:277-280

[16] Bouali I, Trabelsi H, Herchi W, Martine L, Albouchi A, Bouzaien G, et al. Analysis of pecan nut (Carya illinoinensis) unsaponifiable fraction. Effect of ripening stage on phytosterols and phytostanols composition. Food Chemistry. 2014;**164**:309-316

[17] Yang J, Zhou F, Xiong L, Mao S, Hu Y. Lu, B. Comparison of phenolic compounds, tocopherols, phytosterols and antioxidant potential in Zhejiang pecan [Carya cathayensis] at different stir-frying steps. LWT – Food Science and Technology. 2015;**62**:541-548

[18] Prabhakar H, Sharma S, Kong F. Effects of postharvest handling and storage on pecan quality. Food Reviews International. 2022;**38**:1485-1512

[19] Pérez-Fernández V, Spagnoli M, Rocco A, Aturki Z, Sciubba F, De Salvador FR, et al. Non-aqueous reversed-phase liquid-chromatography of tocopherols and tocotrienols and their mass spectrometric quantification in pecan nuts. Journal of Food Composition and Analysis. 2017;**64**:171-180

[20] Robbins KS, Gong Y, Wells ML, Greenspan P. Investigation of the antioxidant capacity and phenolic constituents of U.S. pecans. Journal of Functional Foods. 2015;**15**:11-22

[21] do Pinheiro Prado AC, Manion BA, Seetharaman K, Deschamps FC, Barrera Arellano D, Block JM. Relationship between antioxidant properties and chemical composition of the oil and the shell of pecan nuts [Caryaillinoinensis (Wangenh) C. Koch]. Industrial Crops and Products. 2013;**45**:64-73

[22] Whitfield FB, Mottram DS. Volatiles from interactions of Maillard reactions and lipids. Critical Reviews in Food Science and Nutrition. 1992;**31**:1-58

[23] Erickson MC, Santerre CR, Malingre ME. Oxidative stability in raw and roasted Pecans. Chemical physical and sensory measurements. Journal of Food Science. 1994;**59**:1234-1238

[24] Jiang S, Wang J. Internal quality detection of Chinese pecans (Carya cathayensis) during storage using electronic nose responses combined with physicochemical methods. Postharvest Biology and Technology. 2016;**118**:17-25

[25] Ribeiro SR, Ribeiro QM, Klein B, dos Duarte Santos I, Forgiarini S, Hamann JJ, et al. Effect of low oxygen on quality attributes of 'Barton' pecan nuts after long-term storage at different temperatures. Scientia Horticulturae. 2020;**263**:109098

[26] Prabhakar H, Bock CH, Kerr WL, Kong F. Pecan color change during storage: Kinetics and Modeling of the Processes. Current Research in Food Science. 2022;**5**:261-271

[27] Morales MT, Rios JJ, Aparicio R. Changes in the volatile composition of virgin olive oil during oxidation: Flavors and off-flavors. Journal of Agricultural and Food Chemistry. 1997;**45**:2666-2673

[28] Oro T, Ogliari PJ. Evaluación de la calidad durante el almacenamiento de nueces Pecán [Carya illinoinensis (Wangenh.) C. Koch] acondicionadas en diferentes envases. Grasas y Aceites. 2008;**59**:132-138

[29] Baldwin EA, Wood B. Use of edible coating to preserve pecans at room temperature. HortScience. 2006;**41**:188-192

[30] Thewes FL, Both V, Thewes FR, Brackmann A, Wagner R, Ribeiro SR, et al. Pecan storage: Effects of 1-MCP on the overall quality and volatile compounds profile of shelled and unshelled pecans. LWT. 2021;**145**:111298

Chapter 5

Plant Diseases of Areca Nut

Diannita Harahap

Abstract

Areca nut has source of complex carbohydrates and natural fiber for humans. Almost all parts of this plant can be utilized. This plant is a commodity with high economic value in Indonesia. Moreover, the daily consumption of areca nut is passed down from generation to generation, a tradition is known as 'chew *Pinang*'. The seeds of this plant are always present in the completeness of several traditional events in Indonesia. The development of this plant encountered challenges such as decreased productivity levels due to infection with pathogenic microorganisms. Pathogens that attack areca nut commodities can be in the form of bacteria and fungi. Leaf midrib rot is caused by the mycelia of the pathogenic fungus *Athelia rolfsii*, in addition to infection with *Penicillium* sp. during plant growth also results in losses. *Phytophthora arecae*, Micoplasm-like Organism (MLO), *Ganoderma lucidu*, *Colletotrichum gloeosporioides*, *Curvularia* sp., *Pestalotia palmarum* Cooke., *Cephaleuros* sp., *Ganoderma lucidum*, *Fusarium* sp., *Rhizoctonia* sp., *Xanthomonas campestris*, *Ceralostomelia paradox.*

Keywords: areca nut, Indonesia, microorganism, pathogen, plant diseases

1. Introduction

Areca nut belongs to the category of straight trunked trees with a height of up to 30 meters and a diameter ranging from 25 to 40 meters. The average life span is up to 60 years and some species can reach 100 years. This plant is estimated to be native to Sulawesi (Celebes) Indonesia, Malaysia, and New Guinea [1]. It's distribution covers the tropics in Asia, the Pacific, and parts of East Africa. Areca nut is known by the names beter palm nut, catechu, areca nut, and many other names. This plant can adapt to disturbed land, swamp forest edges, and creeks.

Almost all parts of this plant are economically viable for humans. Plant morphology Areca nut has pinnate leaves with a length of 1-1.5 meters. The base of the midrib leaf is gray and has a crown shape. Branched flower stalks reach a length of 1 meter and grow from under the crown stalk. The fruit has red or orange seeds with a length of 5-6 cm and a width of 4-5 cm. The shape of the fruit varies from round to slightly flat. Taxonomically, Pinang is included in the Spermatophyta division, Monocotyledone class, Order Arecales, Family Arecaceae, and Genus Areca [2].

Areca nut is known as a plant with benefits in every organ it has. These benefits include areas in human life and their role in the ecosystem. The presence of several bacterial phyla (*Proteobacteria* and *Actinobacteria*) in this plant's rhizosphere is known to synergize to help maintain the balance of nutrient cycles in the soil around the roots, thereby maintaining overall betel nut health and increasing plant productivity. This is related to the production of growth hormone, the enzyme 1-Aminocyclopropane-1-Carboxylate (ACC) deaminase which triggers growth and activity against plant pathogens. On the other hand, the use of this plant tree as a yard and garden plant [3]. The stems of this traditional community in Indonesia are used as an important ingredient in the areca climbing competition on Indonesia's independence day by every region in this country. The tradition of the fishing community throws betel leaves into the sea to invite fish to come. In the field of textile dye development, areca nut contributes as a natural dye for Sasirangan batik in South Kalimantan, Indonesia [4]. Areca nut is also reused as ornamental lighting commodities which have a selling value [5]. Young betel nut juice is commonly consumed by the public. Processed smoked betel nut becomes a light snack and is usually used as souvenirs typical of districts in Aceh, Indonesia. In addition, the cosmetic industry has also used betel nut powder as an additional ingredient in the production of transparent soap [6]. Its role in the ecosystem includes a part of agroforestry as an improvement in economic, ecological, and social systems [7]. This cultivation system makes it possible to grow seasonal crops together with crops that can be used in the livestock sector [8]. In addition, areca nut is used for erosion control (dune stabilization) and soil improvement [9].

Based on the description above, the productivity of this plant is very important to be maintained. Disease disturbance in plants is a problem that should be found a solution. Some of the symptoms of areca nut disease and matters that are closely related to the factors that cause the severity of the disease are discussed in this chapter.

2. Factor causing plant disease

Plant health can be a factor that supports plant productivity. A simple definition of a plant health disorder can mean an imbalance in the proportion of normal growth and development caused by biotic and abiotic agents. Abiotic factors such as temperature, humidity, gas atmosphere, and light intensity that provide access to the presence of plant pathogens, decrease productivity or cause plant death directly. Several factors above interact with each other in achieving a state of balance in achieving plant health. Abiotic agents that cause disease in this case are environmental stress conditions and their handling and cultivation. While the biotic agents in question are plant pathogenic infections, including bacteria, fungi, and viruses.

2.1 Abiotic factors

Season (rainy and hot in Indonesia), summer lasts a chance to make the roots dry out, morphologically the roots will shrivel and die. This condition is caused by not allowing the roots to function in absorbing water from the soil.

Low humidity increases evaporation in plants so that moisture stress can occur. When the water supply to the leaves is inadequate, the concentration of the hormone abscisic acid increases rapidly, causing the leaf stomata to close. This process is known to protect plants from excessive water loss. Low humidity in outdoor situations is not usually a problem. This condition is often temporary and rarely results in reports of serious damage. However, if this condition is combined with other factors and occurs at the same time then symptoms of plant disease can appear. These symptoms are temporary to permanent wilting of plants, burnt leaves, and shriveled/wrinkled fruit due to excessive loss of water content. High humidity allows evaporation to occur in plants, inhibiting nutrients from being absorbed by plants. Impaired absorption of nutrients can have an impact on cell formation disorders. For example, plants with young fruit skins are prone to cracking during periods of prolonged high humidity. This provides an opportunity for further pathogens to infect exposed fruit parts without skin dressing.

The temperature fluctuates, while plant physiological activities such as photosynthesis, respiration, membrane permeability, absorption of water and nutrients, transpiration, enzyme activity, and protein denaturation can be directly affected by temperature. The response of plants to the minimum, optimum and maximum temperatures can be different for each plant. Likewise in plants in different places in the same physiological process. In general, plants that grow at the upper and lower limits of the optimum range tend not to grow as well as plants that grow at the optimum temperature range. Sexual organs such as fruit and flowers will grow smaller than plants that grow at optimum temperatures.

The lowest temperatures can cause damage to plant tissue, the level of damage will be exacerbated by stable low temperatures for a long time. Low temperatures can damage the young meristem webs, damaging the tips of young roots. The leaves of the plant will appear yellow and fall. Plant cells can be damaged because low temperatures disrupt the stability of the cell membrane to damage the entire contents of the cell. From low temperatures to freezing points, ice can also damage cell membranes and other vital materials in plant cells. High temperatures can cause damage in the form of coagulation and denaturation of plant proteins, disruption of cell membranes, and secretion of toxins into cells, resulting in cell death. This is through a slowed-down protein synthesis process, the production of 'heat shock proteins to defend the plant from damage. The severity depends on how high the temperature is and the length of exposure.

Wind speed factors affect the spread of plant diseases caused by fungi. This is related to the carrier and spread of conidium and the ability to stick to the plant surface. Light intensity affects several things such as exudation, sporulation, and the dispersion and growth of fungal conidium in the soil. Basically, with adequate light intensity, plant diseases can spread quickly, especially plant diseases caused by fungi. This is supported by related research explaining that conidium *Cercospora* in tobacco can spread and cause picking disease [10]. It also includes environmental imbalances such as air pollution (ozone, hydrogen fluoride, sulfur dioxide, ethylene, and peroxyacetyl nitrate).

2.2 Biotic factors

Pathogenic infections such as bacteria, fungi, and viral pathogens can cause the severity of symptoms that appear and play an important role in the spread of areca nut disease. An explanation of these biotic factors will be discussed in the next section.

2.2.1 Plant disease symptoms

We often observe sick plants in the field. To facilitate sample collection activities, field observation methods are needed. The methods used in field observations on symptoms of plant disease include:

a. Absolute method, observation of plant disease symptoms was carried out thoroughly by the population in the field. The resulting data will represent population estimates in numbers and individuals per unit of the land surface.

b. Relative method, in this condition, the observer will determine a sample of the existing population by considering the sample taken represents the actual conditions in the population. Observing plant disease symptoms in this way will be more difficult to represent the condition of land surface unit conversion.

c. Population index method, observers only measure the indicated part of the attack, it can be a symptom of an attack, and the results obtained are the extent of the attack.

If in the field they have found a method to observe plant disease symptoms, of course, the next step is researchers must be skilled in determining plant disease assessment techniques, here are some things that need to be considered:

i. severity,

ii. describe and compare the symptoms and severity of the disease,

iii. scoring of symptoms,

iv. calculate the attack rate or disease index.

The description below will describe some of the diseases in Pinang caused by. microorganisms in the tropics and subtropics:

2.2.1.1 Symptoms of fruit rot in areca nut

This disease is caused by the fungus *Phytophthora arecae*. This incident began with symptoms of rot and the fall of young fruit from the base of the tree. Symptoms begin with the discharge of dark green/yellowish liquid on the fruit near the petals. The damage spreads to cover the entire surface of the fruit, causing rot. The surface of the fruit is covered with a white mycelium sheath followed by the condition of the fruit stalk rotting and/or dry and not infrequently also covered with white mycelium (**Figure 1**).

2.2.1.2 Symptoms of yellow leaf in areca nut

This disease was discovered as an investment of a Micoplasm-like Organism (MLO) attack. Symptoms are marked by yellowing of the leaf tips and as many as 2-3 strands

Figure 1.
Fruit rot on the areca nut [11].

of branching. Signs of necrosis in the form of brown parallel lines on the leaves, then yellowing extends along the leaf blade and the tips of the leaves dry up. At an advanced stage, the leaves turn yellow and fall off the stem. Symptoms can also be observed at the tips of the roots which gradually turn black. This infection in mature leaves causes the functional leaf cross-sectional area for photosynthesis to decrease (**Figure 2**).

2.2.1.3 Symptoms of foot rot in areca nut

This disease is caused by infection with the fungus *Ganoderma lucidum*. The leaves turn yellow and eventually fall. The stem becomes brittle due to strong winds, and the color of the base of the stem shows a change to brown and secretes a dark liquid. Mushroom hood-shaped fruit grows at the base of the stem, changes in color at the roots and the condition of the roots is drier than before. When the infected stem is cut longitudinally, a brown discoloration can be seen along one meter of the height of the stem from the ground (**Figure 3**).

Figure 2.
Symptoms of yellow leaf disease [11].

Figure 3.
(a) the hood of Ganoderma lucidum *grows at the base of the stem, (b) disease symptoms on the canopy tree [12].*

2.2.1.4 Symptom of bud rot in areca nut

This disease is caused by the fungus *Phytophthora arecae*. Early symptoms that can be observed are leaf changes from green to yellow to brown. The leaves and shoots that appear rot and one by one the leaves fall from the stem (**Figure 4**).

2.2.1.5 Symptoms disease of inflorescence die back in areca nut

This disease is caused by a fungal infection *Colletotrichum gloeosporioides*. Symptoms appear as brownish spots and spread from the tip to the bottom of the inflorescence and cause wilting of the male flowers. Conidia appear as concentric circles in areas of discoloration (**Figure 5**).

2.2.1.6 Symptoms disease of band in areca nut

Symptoms can be observed in the shape of the leaves that are shortened and resemble the shape of a broom, it is not uncommon to find some signs of discoloration to pale wilting and/or yellowish or dark green spots appear. In addition, changes in the

Figure 4.
(A-B) symptoms of leaf color change to yellow then brown, (C) colony morphology Phytoptora arecae *on agar media, (D) conidia of* P. arecae *[11].*

Figure 5.
(a) Brown spots that appear as a symptom disease, (b) wilting as a disease investment [13].

shape of the tree crown such as the shape of a rose, shorten the distance between the trunk segments.

2.2.1.7 Symptoms of yellow spot disease in areca nut

These symptoms appear due to the attack of the fungus *Curvularia* sp. Several symptoms can be observed on the leaves including yellowish spots on the leaf blade, yellow and green spots, the size of the leaves becomes shorter and shaped like a broom, and sometimes the color of the leaves becomes pale, yellow, and drooping. Leaf tips become brown or yellow (**Figure 6**).

2.2.1.8 Symptoms disease of leaf blight in areca nut

This disease is caused by the fungus *Pestalotia palmarum* Cooke. The symptoms of this disease show yellowish spots on the leaf blade, the color of the leaves becomes pale, the tips of the leaves become pale and wilted, and the leaves are shortened and shaped like a broom (**Figure 7**).

2.2.1.9 Symptoms disease of red rust in areca nut

Areca red rust disease is caused by infection with the fungus *Cephaleuros* sp. This disease is characterized by symptoms of yellowish spots, the crown of the leaves withers suddenly, the leaf stalk changes its color to yellowish and the leaf midrib leaves only stick (**Figure 8**).

2.2.1.10 Symptoms disease of root rot in areca nut

This disease is caused by several pathogenic microbes. Some of them are *Ganoderma lucidum* mushrooms. In addition, fungi *Fusarium* sp., and *Rhizoctonia* sp. It also plays an important role in the severity of disease symptoms. Symptoms of this disease show yellowish-brown discoloration and root rot while the plant stem tissue secretes a dark brown liquid. The leaves shorten and resemble a broom (**Figure 9**).

Figure 6.
(A-B) symptoms of leaf spot disease, (C) colony morphology of Curvularia *sp., (D) Conidia* Curvularia *sp. [11].*

2.2.1.11 Symptoms of shoot wilt in areca nut

The crown leaves wilt suddenly, the color of the leaves becomes dull, the tips of the leaves turn brown, and the leaf midrib hangs and finally falls.

2.2.1.12 Symptoms disease of bacterial leaf stripe in areca nut

The fungus *Xanthomonas campestris* Areca is known to be present in areca nut infections. Symptoms of the disease arise in the form of yellowish or dark green spots on the leaf blade, the underside of the leaf surface is covered by bacterial colonies, and yellowish leaf twigs, in the end, the leaf arrangement grows irregularly and is grayish white. The crown of the tree resembles a rose.

2.2.1.13 Symptoms disease of stem bleeding in areca nut

Symptoms of this disease can be observed in stems that change color to dark red, stem tissue secretes dark brown fluid, yellowish spots on leaf blades, roots turn yellowish brown and rot is not uncommon. Another symptom of the fruit shows a change in color

Figure 7.
Symptoms of leaf blight on areca nut: (A) symptoms of brown spot on leaves, (B) colony morphology of Pestalotia palmarum *Cooke., (C) conidia of* P. palmarum *[11].*

to dark green. This disease is known to be caused by the fungus *Thielaviopsis paradoxa*. Von Hohn (*Ceralostomelia paradox*).

2.2.1.14 Symptoms disease of nut splitting in areca nut

Symptoms of the disease that can be observed from this condition include yellowish spots on the leaf blade, yellowish fruit color in the half-ripe fruit phase, and the apical end of the fruit showing signs of falling fruit.

2.2.1.15 Symptoms disease of Antracnose in areca nut

Symptoms that can be observed from this disease are brown circular spots with yellow circles surrounding the spots. The leaf veins turn yellow from the tip to the base of the leaf (**Figure 10**).

2.2.1.16 Symptoms disease of leaf sheath rot

Symptoms that can be observed in midrib rot disease are yellowish white and brown sclerotia on the leaves and covered with white mycelia on the surface. There are first reports of the attack of this disease by the fungus *Athelia rolfsii* in India [14].

Figure 8.
Symptoms of red rust disease on areca nut: (A-B) symptoms of rust spots red on leaves, (C) sign of hyphae growth on leaves, (D) microscopic morphology of Cephaleuros *sp. [11].*

3. Pathogen infection cycle and chain

The disease cycle is inseparable from a pathogenesis [15]. An infection cycle can occur over and over again during a plant's growth period. If this condition continues repeatedly, it is referred to as a chain of infection, so that the pathogen spreads to fill the planting area. The sequence of disease occurrence and disease distribution includes the stages of spore germination, penetration, colonization, infection, sporulation, and disease dispersion. Biochemical processes begin to take place as long as the spores germinate. If the spores are active, if they get the right conditions (host plants and a supportive physical and chemical environment), infection will likely occur. Spores are referred to as dispersion units. Spores germinate morphologically visible germination tube, swelling of the germination tube is known as appressorium, and hyphae to attach when penetration occurs in the host plant tissue. This development becomes the unit of infection. Units of infection can cause disease symptoms [16]. In the areca nut itself, disease symptoms can occur in the vegetative organs and generative organs. Symptoms of the disease can be recognized locally or systemically. If the symptoms occur locally, they only occur in one organ, for example in the leaves. However, if the symptoms are systemic, some organs change, for example, stem bleeding is found in the cross-section of blackish brown fluid, yellowish spots on the leaves, and root rot due to a manifestation of systemic infection.

Figure 9.
Symptoms of root rot disease on areca nut; (A) infection of the base of the stem, (B) systemic symptoms on the tree, (C) fruiting bodies of Ganoderma lucidum, *and (D) spores of* G. lucidum *[11].*

The chain of infection can be divided into homogeneous chains if the dispersion unit consists of only one type. For example, the dispersion unit is only conidia. However, in the case of an infection chain consisting of several dispersion units such as ascospores and conidiospores, it is said to be a heterogeneous infection chain. Very often this heterogeneous chain of infection occurs in different seasons in perennial plants [16].

4. Plant defense response

The defense response of plant cells against pathogenic microbial infections are divided into two mechanisms, including mechanisms related to 1) the structure of the plant body itself including the presence of a waxy layer on the surface of plant organs. On the other hand, hard and thick epidermal cells narrow the chance of pathogen penetration. In addition, plant defenses against pathogens are also found in the various structures of stomata and guard cells as natural access that plants have to pathogens. For many plant varieties with narrow entrance variations and tall and wide guard cells reduce the chances of pathogens penetrating; 2) production of antimicrobial secondary metabolites. This is a limiting factor for the continued penetration of pathogenic microbes [16].

When pathogens attack plant cells, plant enzymes catalyze oxidative reactions to produce reactive oxygen species that can damage pathogenic microbial cells.

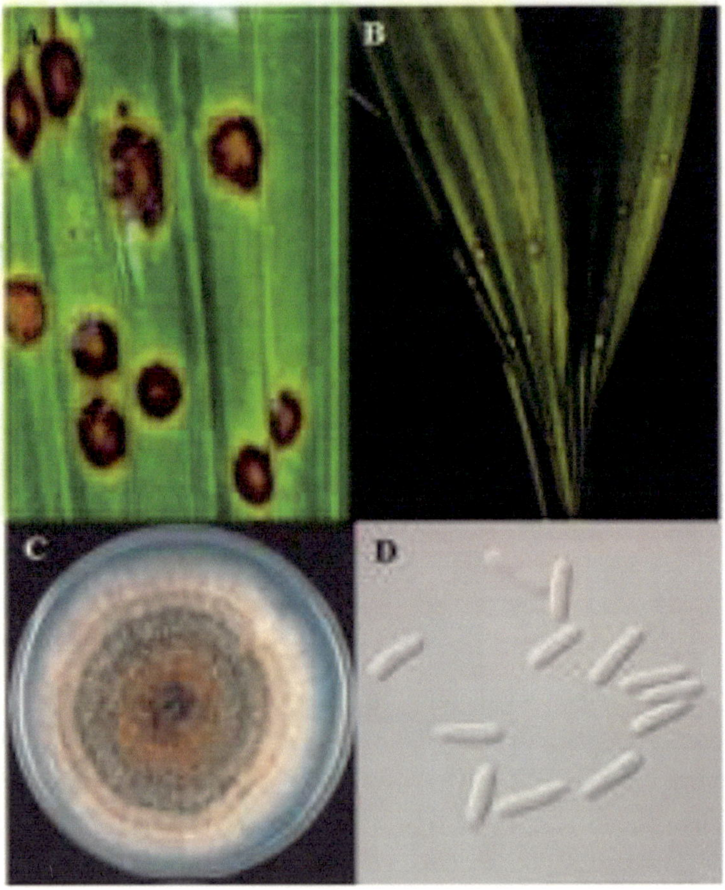

Figure 10.
Antracnose; (A-B) symptoms of brown spots on leaves with a yellow halo around them, (C) morphology of the fungus Colletotrichum gloeosporioides, *(D) conidia of fungi* C. gloeosporioides *[11].*

This reactive oxygen also serves to strengthen cell walls as well as signal intermediaries to neighboring cells that foreign cell attacks are taking place. In addition, plant cells respond by synthesizing and storing some callose (polysaccharide polymer) which is located between the cell wall and the plant cell membrane. Callose blocks the penetration of pathogenic cells at the site of infection [16].

5. Genetic variations of areca plants in Indonesia and resistance to pathogens

Diseases of areca nut attack the vegetative and generative organs. Vegetative organs that are the target of disease-causing pathogens include roots, stems, and leaves. Meanwhile, generative organs such as seeds, fruits, and flowers cannot be separated from the attack of pathogenic infections (**Table 1**).

No.	Accesion	Island Origin in Indonesia	References
1	Sumut-1, Sumbar-1, Sumbar-2, Sumut-2, Bengkulu-1, Sumbar-3, Bengkulu-2, Galang Suka (GSK), Tarean (TRN), Betara-1, Betara-2, Muara Sabak Timur-1, Muara Sabak Timur-2, Muara Sabak Timur-3, Wangi, lokal	Sumatra	[17, 18]
2	Molinow-1 (MLW-I), Molinow-2 (MLW-II), Mongkonai (MGK), Duhia Da'a, Tingkohubu I, Tingkohubu II, Huntu I, Huntu II, Huntu III, Bulawan	Sulawesi	[19]
3	Oyehe (OYH), Kali Harapan (KHN), Nifasi (NFS)	Papua	[20]
4	Saluang	Kalimantan	[21]

Table 1.
Areca nut varieties in Indonesia.

New varieties are obtained from studying genetic diversity. One way to determine the diversity of a betel nut species is based on differences in morphological characteristics (either qualitatively or quantitatively) which are directly related to fruit production. Genetic diversity can be obtained by crossing, introduction, mutation, and selection of genetic sources. Selection based on morphological characters is a method that is often done. The selection was made on the morphological characteristics of the areca nut that were directly related to fruit production, namely the number of bunches, the number of fruit marks, and the number of spikelets [22]. Some of the morphological characteristics of areca nut that are often observed in genetic diversity research include stem height, stem circumference, number of leaf marks, number of leaves, number of bunches, number of fruit per bunches, whole fruit weight, fruit polar length, fruit equatorial length, fruit kernel weight, kernel polar length and kernel equatorial length [23]. One of the Pinang varieties in Indonesia which is classified as a superior variety is Pinang Betara. This variety received a recommendation from the government in the Decree of the Minister of Agriculture No. 199/Kpts/SR.120/1/2013. The characteristics of Pinang Betara include 1) dry kernel production/tree/year 5.70 kg, stem height reaches 10 meters, has 131.35 fruit/bunch, whole fruit weight is 47.06 grams, dry seed weight is 8.68 grams, high 9.79% tannin, has a dry kernel potential of 7.81 tons/hectare (**Figure 11**) [24].

The incidence of root rot can be caused by plant genetic factors so that the resistance response of each plant variety can be different [25]. Different genetic factors between plant varieties are derived from different inherited genes. Each variety has a different resistance to *S. rolfsii* fungus attack. The resistance properties found in each of these varieties are coded for resistance genes to fight pathogens. In other words, the resistance control genes possessed by each variety are different [26]. The Character of resistance to *Fusarium* sp. wilt disease. Judged by several things: 1) disease period value (day after inoculation), 2) disease incidence (%) and 3) endurance category (units/day). According to this study, a longer incubation period correlates with lower disease incidence and slower infection rates [27].

Figure 11.
Several varieties of areca nut in Indonesia, from Sumatra Island: (A) Pinang Wangi, (B) Pinang lokal [23], (C) Pinang Betara [22], from Sulawesi island: (D) Pinang Duhia Da'a-Gorontalo, Tingkohubu, Huntu I, Huntu II, Huntu III [19].

6. Conclusion

Diseases in plants can be caused by an imbalance of environmental conditions, pathogens with high pathogenicity, and susceptible hosts. These three factors trigger disease severity in plants. This paper describes sixteen diseases in areca nut caused by the presence of pathogens, infection cycle and chain, plant defense response, as well as genetic variation affects plant resistance to pathogens.

Acknowledgements

The author would like to thank the head of the biology department, the Dean of the Faculty of Science and Technology, and The Rector of Universitas Islam Negeri Ar-Raniry Banda Aceh, Indonesia who has allowed the author to contribute to this chapter of the book.

Conflict of interest

The authors declare no conflict of interest.

Author details

Diannita Harahap
Faculty of Science and Technology, Department of Biology, Universitas Islam Negeri Ar-Raniry, Banda Aceh, Indonesia

*Address all correspondence to: diannitaharahap@ar-raniry.ac.id

References

[1] Heatubun CD, Dransfield J, Flynn T, Tjitrosoedirdjo SS, Mogea JP, Baker WJ. A monograph of the betel nut palms (areca: Arecaceae) of east Malesia. Botanical Journal of the Linnean Society. 2012;**168**:147-173. DOI: 10.1111/j.1095-8339.2011.01199.x

[2] NCBI. Available from: https://www.ncbi.nlm.nih.gov/Taxonomy/Browser/wwwtax.cgi?id=184783. [Accessed: May 27, 2022]

[3] Sagrim I, Soekamto MH. Pembibitan Tanaman Pinang (Areca catechu) Dengan Menggunakan Berbagai Media Tanam. Median J Ilmu Ilmu Eksakta. 2019;**10**:28-36. DOI: 10.33506/md.v10i2.295

[4] Prabawa IDGP. Ekstrak Biji Buah Pinang Sebagai Pewarna Alami Pada Kain Sasirangan. J Ris Ind Has Hutan. 2014;**7**:31. DOI: 10.24111/jrihh.v7i2.1229

[5] Alvaran MSA, Masri A. Eksplorasi Buah Pinang Merah Sebagai Material Produk Lampu Hias. SERENADE Semin Res Innov Art Des. 2022;**1**:36-43. DOI: 10.21460/serenade.v1i1.10

[6] Yernisa EG-S, Syamsu K. APLIKASI PEWARNA BUBUK ALAMI DARI EKSTRAK BIJI PINANG (Areca catechu L.) PADA PEWARNAAN SABUN TRANSPARAN APPLICATION OF NATURAL DYE POWDER FROM SEEDS OF Areca catechu L. IN TRANSPARENT SOAP Yernisa (1)*, E. Gumbira-Sa'id (2), dan Khaswar Syamsu (2). J Tek Ind Pert. 2013;**23**:190-198

[7] Tiurmasari S, Hilmanto R, Herwanti S. Analisis Vegetasi Dan Tingkat Kesejahteraan Masyarakat Pengelola Agroforestri Di Desa Sumber Agung Kecamatan Kemiling Kota Bandar Lampung (vegetation analysis and prosperous of society agroforestry organizer in the Sumber Agung Village Kemiling District B). J Silva Lestari. 2016;**4**:71-82

[8] Rendra PPR, Sulaksana N, Alam YCSSS. Optimalisasi Pemanfaatan Sistem Agroforestri Sebagai Bentuk Adaptasi Dan Mitigasi Tanah Longsor. Bull Sci Contrib. 2016;**14**:117-126

[9] Darwo. Tanaman Pinang untuk Selamatkan Gambut. 2020. Available from: https://kabaralam.com/tapak/tanam-pinang-untuk-selamatkan-lahan-gambut. [Accessed: August 5, 2022]

[10] Tantawi AR. Hubungan Kecepatan Angin dan Kelembaban Udara terhadap Pemencaran Konidium Cercospora nicotianae pada Tembakau. AGRITROP. 2007;**26**:160-167

[11] Mubin N, Khairani HS, Triwidodo H, Bandi D. Buku Saku Hama dan Penyakit pada Komoditas Jambu Mete, Kelapa, Sirih dan Pinang. Department of Plant Protection, Faculty of Agriculture. IPB University. 2021. p. 185. ISBN 978-602-96419-8-1

[12] Indonesia BNR. Pedoman Teknis Pembangunan Kebun Sumber Benih Pinang. Vol. 1828. Indonesia: Indonesia BNR; 2014

[13] Arecanut Diseases. n.d. Available from: http://www.kissankerala.net:8080/KISSAN-CHDSS/English/index1.html. [Accessed: May 5, 2022]

[14] Kamil D, Bahadur A, Debnath P, Kumari A, Choudhary SP, Prameeladevi T. First report of leaf

sheath rot of areca nut (Areca catechu) caused by athelia rolfsii from Tripura, India. Plant Disease. 2020;**104**:104. DOI: 10.1094/PDIS-01-20-0157-PDN

[15] Sutarman. Dasar-Dasar Ilmu Penyakit Tanaman. Sidoarjo: UMSIDA PRESS; 2017

[16] Suniti NW. Buku Ajar Epidemiologi Penyakit Menular. Denpasar: Jurusan Agroteknologi Fakultas Pertanian Universitas Udayana; 2016

[17] Miftahorrachman M. DIVERSITAS GENETIK TUJUH AKSESI PLASMA NUTFAH PINANG (Areca catechu L.) ASAL PULAU SUMATERA. J Penelit Tanam Ind. 2020;**12**:27. DOI: 10.21082/jlittri.v12n1.2006.27-31

[18] Warta Penelitian. 2012. pdf n.d.

[19] Miftahorrachman MI. KERAGAMAN GENETIK PLASMA NUTFAH PINANG (Areca catechu L.) DI PROPINSI GORONTALO. J Penelit Tanam Ind. 2020;**13**:119. DOI: 10.21082/jlittri.v13n4.2007.119-124

[20] Bagindo M. Keragaman Genetik Aksesi Pinang (Areca catechu L.) Asal Papua, Sulawesi Utara dan Sumatera Utara (Random Amplified Polymorphic DNA). 2011. pp. 1-17. Thesis summary. Available from: http://repository.ipb.ac.id:8080/handle/123456789/51523 [Accessed: August 9, 2022]

[21] Arni. Pinang Saluang, Potensi Lokal Komoditas Ekspor Unggulan Kalbar. n.d. Available from: https://www.suarakalbar.co.id/2022/03/pinang-saluang-potensi-lokal-komoditas-ekspor-unggulan-kalbar/. [Accessed: August 2, 2022]

[22] Miftahorrachman Nur M. Hubungan Produksi Buah Dengan Beberapa Karakter Vegetatif Dan Generatif Populasi Pinang Betara Melalui Analisa Koefisien Lintas Correlation of Production with some Morphology Characters of Betara Arecanut Population by Path Coefficient Analysis. Palma Research Institute; 2013. pp. 34-40

[23] Miftahorrachman N, Mantau F, Hervina D. Evaluasi Karakter Morfologi Untuk Perbaikan Genetik Tanaman Pinang (areca Cathecu L.) di Padang Pariaman [evaluation of morphological characters for Arecanut (areca Cathecu L.) genetic Improvment In Padang Pariaman]. Bul Palma. 2018;**19**:15. DOI: 10.21082/bp.v19n1.2018.15-25

[24] Nasution F. Mengenal Varietas Bibit Unggul Asli Indonesia, Pinang Betara. 2019. Available from: https://www.bibitpinangunggul.com/2019/03/bibit-pinang-unggul-asli-indonesia.html. [Accessed: May 20, 2022]

[25] Yuliani D, Wage D, Rohaeni R. Heritability, gene resource, and durability of Rice varieties resistance to bacterial leaf blight disease. J Litbang Pertan. 2017;**36**:99-108

[26] Fitria MR. Respon ketahanan dan kandungan senyawa fenol enam varietas kedelai (Glycine max (L.) Merrill) terhadap penyakit busuk pangkal batang (Sclerotium rolfsii Sacc.). Berk Ilm Pertan. 2020;**3**:27-32

[27] Aprilia I, Maharijaya A, Wiyono S. Keragaman Genetik dan Ketahanan terhadap Penyakit Layu Fusarium (Fusarium oxysporum f.sp cepae) Bawang Merah (Allium cepa L. var. aggregatum) Indonesia. J Hortik Indones. 2020;**11**:32-40. DOI: 10.29244/jhi.11.1.32-40

Chapter 6

This is the Nut You Should Be Eating for Better Gut Health

Giorgio Calabrese and Cinzia Myriam Calabrese

Abstract

The consumption of dried fruit is a dietary habit that is growing year by year. Nutrition specialists and others have carried out numerous studies confirming that the introduction of a controlled daily intake in the diet can bring benefits both in the prevention of major metabolic and cardiovascular diseases and in the well-being of the microbiome. The intake of dried fruit plays an important role in improving the quality and quantity of the microbial families of our intestine, leading to an increased production of short-chain fatty acids, which participate in numerous pathophysiological mechanisms and in the prevention of numerous diseases. The present review chapter compares numerous articles focusing on the characteristics and behavior of fibers and other main components of nuts.

Keywords: dried fruit, nut, Mediterranean diet, microbiome

1. Introduction

Oily dried fruit is a different food from dried fruit (figs, apricots, plums, and raisins) because the latter has a high sugar content and is therefore particularly suitable for the beginning of the day because it provides a load of immediately available energy. Dried fruit, such as nuts and various seeds, is the vehicle for the reproduction of vegetables; within them is the potential of the entire plant. The term "walnut" commonly indicates the seed of a tree enclosed in a woody casing, but many seeds that we consider "nuts," such as peanuts, do not correspond to the strict definition of pure walnut.

There are more than 300 types of walnut, and all of their plants produce seeds, but only a few are very important as a product for food and human health

In terms of world production, coconut is the most widely grown and used, followed by peanuts; both account for about 94% of world walnut production. The main reason is that both coconut and peanuts provide oils that are among the essential ingredients of cooking oils, margarine, and cooking fats at a low price. For example, peanuts are the most known nut in the United States, and the harvest exceeds 70% of the annual nut production, followed by almonds, walnuts, and pecans.

Unfortunately, nuts are usually eaten after being fried in fat and salted or as an ingredient in sweets and packaged and ultra-processed foods. Today, many people are trying to orient themselves to healthier foods, and the consumption of nuts and oil seeds is growing dramatically!

It is worth mentioning that oily fruit contains essential nutrients and substances defined as "anti-nutrients" but always to a lesser extent than healthy ones, which are fibrous compounds, protease inhibitors, ellagic acid, and other polyphenols. Since nuts and seeds contain a large amount of oil, one might think that eating nuts frequently would increase the rate of obesity. Still, worldwide epidemiology shows us, on the contrary, that obesity is less common among nut eaters, and one of the positives is thought to come from a greater sense of satiety and appetite satisfaction. In general, nuts and oil seeds, due to their high oil content, should be purchased and stored still in the shell because it is a natural protector against free radical damage caused by light and air. It is discouraged to eat moldy nuts and seeds and use them for cooking because they pose a health risk, just as the use of chewy, dark, or wilted nuts is similarly not recommended. Instead, it is advisable to store nuts and oilseeds in their shells in a cold, dry environment, and if by chance you are forced to buy them unshelled, be sure to seal them in airtight containers in the refrigerator or freezer. Chopped or sliced walnuts, in general, are rancid, and it is preferable to prepare them at home, using whole walnuts.

The nutritional characteristics of the two types of fruit are a high sugar content for dried fruit and a high quantity of calories for dried fruit. The high-calorie intake is due to its high fat content, especially polyunsaturated fatty acids, such as alpha-linolenic acid (ALA). Another type of fat found in this food is eicosapentaenoic acid (EPA). These two types of fats are an excellent source for people who do not eat fish.

Oily dried fruit is a large family, including hazelnut, with high fat content. It is a food that is present in diets because its characteristic aroma makes it versatile in the combination of other flavors, and it is a widely used ingredient in commercial confectionery products. During the second world war, cocoa was challenging to find, and hazelnut was used to prepare products that resembled chocolate. After the war, when the cocoa supply returned to normal, confectionery producers found that the public's taste remained oriented toward specialties in which the hazelnut aroma sweetened the cocoa aroma's sourness. Consumers' preferences are generally oriented toward large, somewhat round, well-colored, and hard-shelled hazelnuts. Hazelnuts contain omega-3 and omega-6 unsaturated fatty acids and are rich in fibers that facilitate intestinal transit and, in turn, make you absorb less sugar and less fat.

Walnuts contain the kernel, the part eaten with a pleasant flavor and with a delicate taste. Walnut is very nutritious and has a high calorific value, as it includes a characteristic oil in a percentage that varies, depending on the variety, from 55% to 65%. The high-calorie nature of walnuts makes them undesirable for those who want to lose weight, as they provide many calories.

Walnut oil contains alpha-linolenic acid (ALA), a precursor of omega-3 fatty acid, called EPA. The intake of four walnuts per day leads to the presence of EPA in the body, similar to that obtained from eating fish. The nut is appreciated as a dessert and is sought after as a delicacy at the end of a meal, as it stimulates the taste of drinking.

From a nutritional point of view, almonds are similar to walnuts. They, too, contain an oil with a high percentage of omega-3 and omega-6 and a minimum amount of water; almond is, therefore, a high-calorie food. It is an essential source of mineral salts such as phosphorus, calcium, iron, copper, and magnesium. Almonds are often eaten dry and can be distinguished as dry and bitter. They can hardly be digestible due to their high oil content.

Almond oil is extracted by applying pressure to a sweet variant. When it is emulsified with water and sugar, a natural laxative is obtained that can also be given to children. If you add a small dose of bitter almond oil to this preparation and dilute it, it becomes the basis for preparing barley.

For many years, numerous scientific studies have confirmed that a balanced diet, such as the Mediterranean diet, allows you to prevent numerous chronic diseases and that, in pathological conditions, diet is an integral part of medical therapy. Diet plays a crucial role in the etiology of numerous chronic diseases, thus contributing to significant geographic variations in chronic disease morbidity and mortality rates in different countries and populations worldwide [1].

The daily eating style, in addition to having an essential role in prevention, can be one of the causes of some metabolic diseases if you follow an unbalanced diet. It has also been found that it significantly influences the composition of the intestinal microbiota [2], where we can distinguish about 1014 commensal bacteria [3] grouped into seven different divisions: the most numerous are those of the Bacterioidetes families (in particular, Bacteroides and Prevotella) and Firmicutes, which are more than 90% of the total population [4], and in the remaining 10%, we find archaea, viruses, parasites, and fungi [5].

The microbiological families make up the microbiota and feed on the fermentation of non-digestible food carbohydrates [6], producing gas, some organic acids, and short-chain fatty acids (SCFA) [4, 6, 7]. The latter are acetate, propionate, and butyrate and have a crucial role in human health [7]. Butyrate acts on the intestinal mucosa by activating the synthesis of mucin and favoring the formation of tight junctions [8], essential for maintaining intestinal permeability.

SCFA, as confirmed by the scientific literature, plays a role in the modulation of the immune system [9, 10] by regulating the production of pro-inflammatory cytokines and chemokines [9, 11] and the function of T lymphocytes [10].

2. What is a dried fruit?

When we talk about taking dried fruit in the daily diet, we refer to nuts, including oily achenes.

Walnut is one of the many names used to refer to the oily achenes produced by different plant species. These include walnuts, hazelnuts, almonds, pistachios, pine nuts, peanuts, Brazil nuts, pecans, cashews, and macadamias. Generally rich in protein and exceptionally energizing, the above-mentioned oily nuts represent an essential natural nutritional choice in vegetarian and sports nutrition. Thanks to their richness in fiber and excellent content of mono- and polyunsaturated fatty acids (especially walnuts), they exert a protective action against the so-called wellness diseases, such as diabetes, hypercholesterolemia, and obesity.

For these assumptions to be valid, dried fruit must be consumed sparingly, replacing (never adding to) other less-healthy lipid sources (such as animal fats). Given its high caloric value and great fat richness, dried fruit should be eaten away from main meals, contextualized in snacks, perhaps in combination with fresh fruit (e.g., an apple and some almonds or a kiwi or some walnuts).

A reasonable serving shouldn't exceed 30 g, however. If associated with fresh fruit, it represents an excellent snack and a good ally for our health and daily well-being. The dried fruit marketed with its shell does not guarantee the quality of the product, but it is healthier than a chopped or shelled one. The latter category of food, even if packaged in sealed containers, requires the addition of artificial antioxidants, necessary to avoid the rancidity of fats and extend storage time.

The scientific communities have conducted and are conducting scientific studies for a healthy diet, where the daily intake of dried fruit is recommended. In one of the most significant studies on the Mediterranean diet PREDIMED enrolled in 2003 and 2011, over 7000 people were at high risk of cardiovascular events. The trial divided the participants into three groups: a Mediterranean diet supplemented with extra virgin olive oil, a Mediterranean diet enriched with dried fruit (walnuts, hazelnuts, and almonds), and a low-fat isocaloric diet (animal and vegetable). The two groups assigned the Mediterranean diet adhered well to the dietary regimen and improved clinical analyses. The study showed that for patients with high CVD risk, the Mediterranean diet with extra virgin olive oil or dried fruit intake reduces the incidence of cardiovascular disease [12].

For some time now, clinical studies have focused on the beneficial properties of dried fruit as cardiovascular prevention, especially for coronary heart disease, and then extended to other pathologies [13].

Among the epidemiological studies carried out over the years, it has been found that a regular intake of dried fruit reduces the risk of death from CVD and a reduction in the risk of developing cardiovascular disease [14].

The benefit of dried fruit is the cholesterol-lowering effect, which explains its role in cardiovascular prevention. One of the most prominent large-scale cohort studies of 2005 showed a significant inverse association between the frequency of nut consumption and inflammatory markers (c-reactive protein, IL-6, and fibrinogen) [15].

Nutrition experts in dietary prescriptions recommend replacing ultra-processed foods that exceed in fat, sugar, and salt with nuts as snacks. This substitution helps lower the diet's glycemic load and therefore reduces the risk of developing cardiovascular disease. Among the various benefits of consuming nuts is the satiating power. The consumed fibers fill and quench hunger, making the next meal less excessive.

Having a low glycemic index, walnut, and other types of nuts, combined with a good amount of fat, reduce the total glycemic load of the meal.

The particles produced contain intact plant cells that derive from their fibrous and indigestible wall during chewing. The plant cell wall reduces the entry of digestive enzymes, slowing digestion and absorption of about 25% of the fats. The abrasive consistency of oily fruits allows these foods to be natural exfoliants, removing plaque and stains from tooth enamel. Once in the stomach, they manage to activate the satiating power. In the intestine, soluble fibers such as inulin help reduce the gases produced physiologically by the intestinal microbiota.

3. The nutritional differences of dried fruit

Nuts are also called lipid fruit, as they are low in water and high in fat. The most common are:

- **Walnuts**: They contain little water (3–4%); a good share of oil (60%); and a share of proteins (20%), vitamin E, calcium, iron, and zinc.

- **Almonds**: They contain many lipids (up to 60%), polyunsaturated fatty acids, proteins (20%), potassium, calcium, iron, and zinc. The amygdalin (laetrile) in them has anticancer activity. There is an excellent relationship between arginine and lysine; therefore, they not recommended for those suffering from herpetic infections.

- **Pine nuts:** They are nutritionally very close to almonds, but they go rancid more easily and therefore require storage in the fridge in hermetically sealed containers.

- **Pistachios:** They are nutritionally similar to almonds, with a higher content of iron and thiamine and a lower content of calcium and niacin. They contain lipids (55%), proteins (20%), and B vitamins.

- **Hazelnuts**: Compared with almonds, they contain more lipids and are less protein rich. They are more caloric; 100 g corresponds to about 634 kcal.

- **Chestnuts:** They contain many more carbohydrates, a good amount of protein, and little fat. They are an excellent source of potassium, magnesium, iron, and manganese.

- **Cashews**: They contain a few lipids, of which 65% is composed of unsaturated fats, such as monounsaturated oleic acid (90%) and linoleic acid (10%). They contain magnesium, potassium, and zinc. The shell of cashews contains caustic oil, which, if in contact with the skin, is harmful.

- **Peanuts**: They contain lipids (50%), proteins (20%), and carbohydrates (19%). They also contain B vitamins, potassium, magnesium, calcium, zinc, and iron. 75% of the fats present is unsaturated.

- **Brazil nuts**: They are similar to almonds, with more lipids. They also contain methionine and cysteine, which are helpful for those who follow a vegan diet.

- **Macadamia nuts:** They contain a lot of fat (72%) and little protein (8%).

- **Pecan nuts**: They have high fat content (71%), low protein, mineral salts, and B vitamins.

The queen of dried fruit is the walnut. The walnut is a very ancient plant; 9000 years ago, people started talking about this fruit. The name derives from the Latin "nux nucis," which included the walnut and similar fruits, those with a fleshy kernel, wooden shell, and fibrous rind.

The *Jungals regia* fruit tree (or white walnut) produces a fruit, the drupe, which is fleshy and composed of the husk (the fibrous outer part), and the shell (endocarp) that is the woody inner part. Inside the shell is the seed (kernel), which is edible. This plant, to date, is cultivated in many parts of the world with different cultivars:

- United States: We find Hartley and Howard nuts.

- California: We find Eureka, Payne, and Chandler.

- France: We find Franquette, Parisienne, Soleze (or Argor), Mayette, Walnut of Grenoble (has the AOP certification), Lara, and Meylannaise.

- Italy: Sorrento walnut, common walnut, Maliza, Corncola, Late or San Giovanni walnut, Bleggiana walnut, and Feltrina walnut.

Chandler is the most cultivated and resistant to infections and diseases among all these cultivars.

The nutritional characteristics per 100 g of walnuts are as follows:

- Energy: 689 kcal
- Carbohydrates: 5.1 g
- Starch: 1.8 g
- Soluble sugars: 3.1 g
- Protein: 14.3 g
- Lipids: 68.1 g
- Water: 3.5 g
- Total fiber: 6.2 g
- Soluble fiber: 0.84 g
- Insoluble fiber: 5.37 g
- Cholesterol: 0
- Calcium: 83 mg
- Potassium: 368 mg
- Sodium: 2 mg
- Phosphorus: 380 mg
- Iron: 2.1 mg
- Magnesium
- Zinc: 2.7 mg
- Copper: 1.5 mg
- Selenium: 3.1 μg
- Thiamine: 0.45 mh
- Riboflavin: 0.1 mg
- Niacin: 1.9 mg

- Vitamin C: traces
- Vitamin A: 8 µg
- Vitamin E: 3 mg

4. Benefits of dried fruit

Nuts are a valuable ally of our heart; their composition, especially of unsaturated fatty acids, helps reduce inflammation, improves blood pressure control, and regulates blood cholesterol levels: phytosterols, fiber, and mono- and polyunsaturated fatty acids help to contain LDL (the cholesterol dangerous for the heart) and to maintain reasonable levels of HDL (the "good" one), reducing the process of atherosclerosis and thus limiting the risk of blockage of the arteries.

In literature, there are many scientific studies with walnut as the protagonist. It is present in many dietary regimes, mainly the Mediterranean diet.

Among its nutritional characteristics, walnut is rich in good fats, especially omega-3 fatty acids, such as alpha-linolenic acid (ALA). The latter is the primary omega-3; it is an "essential" polyunsaturated fatty acid because our body cannot produce it on its own, hence the importance of consuming it with the diet. ALA keeps blood cholesterol levels low.

Walnut is also a source of essential minerals such as iron, magnesium, phosphorus, potassium, zinc, copper, and vitamin E—a powerful antioxidant. It does not contain cholesterol but a minimum amount of saturated fats and an overall percentage of mono- and polyunsaturated fats. This feature allows it to provide energy and carry vitamins (folate, niacin, and vitamin E) and minerals (magnesium and potassium), which are helpful not only for bone health, muscles, heart, and circulation, but also for the brightness of the skin and the strength of the hair.

Mono- and polyunsaturated fats are the components of cell membranes and some hormones, and they help keep the immune system active and the blood thin. They also contain bioactive compounds and protective substances such as tocopherols, phytosterols, and phenolic compounds and proteins and carbohydrates in balanced quantities and lots of fiber.

Given its composition, the walnut turns out to be a high-calorie food, so its consumption must be carefully dosed. An average walnut weighs about 8 g with a caloric intake of 35 kcal. Nutrition specialists recommend an intake of 15 g of walnuts per day for a caloric consumption of approximately 100 kcal.

Walnuts are very versatile in diets; they can be eaten as a snack or as a dessert, but they can also be associated with other foods, making them more palatable.

Given its nutritional characteristics, walnut is not considered only as a fruit or food. The European Commission, in regulation 43/2012, defined walnut as a natural nutraceutical food [16] as it improves the elasticity of blood vessels if consumed with a dose of 30 g per day. The nutraceutical power is given by the number of plant sterols, notably polyphenols, whose biological activity is potent. There are also large quantities of ellagitannins and ellagic acid, with antiviral, antibacterial, and antitumor properties [17].

Comparison with observational studies on European and American populations has shown that the Mediterranean diet, where walnuts are very present, leads to a reduced risk of mortality from numerous diseases over a 10-year follow-up [18].

A report on the PREDIMED study revealed that a group of about 1200 patients who followed the Mediterranean diet with walnut supplementation for 1 year gave a 14% reduction in the prevalence of metabolic syndrome, compared to a 2% reduction in the group assigned low-fat diet [19].

Another cross-sectional study compared four different diet regimens in a multi-ethnic group in North America. The group following a diet with cereals, fruit, and nuts showed a reduced risk of cardiovascular disease and a positive effect on inflammatory markers (inflammatory biomarkers, endothelial function, and homocysteine) [20].

From the scientific evidence, it can be asserted that the consumption of walnuts has a lipid-lowering effect; they reduce the risk of developing diabetes, especially in the female population and are not associated with weight gain and therefore not associated with obesity [21].

A recent meta-analysis evaluated the effect of walnut consumption on all causes of mortality from cardiovascular disease to cancer. The cohort studies were analyzed to support the decrease in mortality in individuals who eat walnuts [22].

Walnuts contain various dietary nutrients such as fiber, vitamins (folic acid, niacin, tocopherols, and vitamin B6), minerals (calcium, magnesium, and potassium), phytosterols, and phenolic compounds [23].

The fats present in walnuts are low in SFA and high in MUFA (oleic acid), with an abundant amount of PUFA (a-linolenic acid, omega 3) [24]. We also find a high content of L-arginine, a precursor of nitric oxide, with vasodilating power [25].

The cholesterol-lowering power is given by the phytosterols present in walnuts [26].

A recent study on the Chinese-Asian population targeting the prevention of type II diabetes mellitus and glycemic changes has considered the postprandial glycemic response (0–120 min) after the administration of high-protein nut bars compared to the administration of high-carbohydrate cereal bars. About 16–28 g of mixed nuts were also given daily. The objective was to evaluate the postprandial response of overweight adults with an increased risk of developing type 2 diabetes. The study showed that the intake of walnut bars suppressed the glycemic response and improved the postprandial hyperglycemic response. The high-risk group observed a 30% reduction in the overall glycemic impact [27].

Previous studies have confirmed the association between walnut intake and decreased cardiovascular risk and the development of type 2 diabetes mellitus [28, 29].

Some previous studies have shown that the intake of nuts suppresses postprandial blood sugar caused by the intake of foods rich in carbohydrates [30, 31].

5. Microbiome and dried fruit

Numerous families of bacteria populate the microbiome present in our intestines [32]. Numerous factors influence microbiota, such as age, diet, drugs, and the environment. These impact bacterial families in metabolizing nutrients derived from our diet and some drugs. The microbiota is a form of barrier for our intestines, it protects us from infections and, at the same time, it produces molecules (proteins, fatty acids, etc.) that participate in numerous physiologicals processes.

When the microbiome is altered, characteristic intestinal disorders such as irritable bowel syndrome (IBS) [33], constipation [34], and abdominal bloating can be correlated [35].

The intake of walnuts is healthy not only to reduce cardiovascular risks but also to promote the balance of intestinal bacteria in our microbiota.

Walnuts can be considered prebiotics and probiotics due to their bacterial content. Their peel contains many fibers. Their fermentation promotes the production of SCFA (such as butyrate, which optimizes intestinal contractility and mucus secretion) [36] and polyphenols with the antioxidant and anti-inflammatory properties.

It can be said that the consumption of dried fruit particularly promotes the growth of the Rosuberia bacterial family, which protects the intestinal lining. The bacterial family Eubacteria Eligens also grows after the intake of walnuts and is associated with an improvement in blood pressure.

The bacterial family of Lachnospiraceae has been associated with more significant reductions in blood pressure, total cholesterol, and non-HDL cholesterol.

In 2020, a systematic review and meta-analysis of controlled studies on the effect of nut intake on the microbiota was published. The study also evaluated bowel function and its symptoms in healthy adults [37].

Studies with findings related to fecal microbiota were included in the review; all studies measured how phyla behaved after walnut intake, in particular Actinobacteria, Bacteroidetes, Firmicutes, Proteobacteria, and Verrucomicrobia [38, 39].

From this comparative study, it can be confirmed that the consumption of walnuts affects the composition of the microbiota. Still, given the results, there is a confirmation at the level of gender and not directly on the species or the diversity, depending on the duration of intake of the walnuts. Although the number of comparative studies is few, the study on the effects of nut consumption on the microbiota needs further trials [40].

In another comparative study [41], the microbiota was evaluated in a pediatric and adult population. All enrolled children had reached the age of three, as a 3-year-old child's microbiota is comparable to an adult's [42]. This population was given whole or partially processed walnuts and compared with populations that either did not eat walnuts or ate them in negligible doses.

The studies measured the microbiota with innovative techniques, with complete analyses allowing for quantification and taxonomic identification.

The Holscher study was compared where the population was given 42 g of walnuts daily for three weeks. No significant differences were found in the size of the entire sample, but the intake of walnuts significantly influenced the bacterial families. A decrease in Actinobacteria and a substantial increase in Firmicutes were observed. An increase in the genera Faecalibacterium, Clostridium, Roseburia (cluster of Clostridium XIVa and IV), and Dialister (49–160% more relative abundance) and a reduction in the genera Ruminococcus, Dorea, Oscillospira, and bifidobacteria (16–38% relative abundance inferior) were also observed [39].

From the data that emerged, it can be deduced that a diet with numerous different foods can temporarily reduce microbial diversity. This consequence is called the "shock effect." These results contradict those in the scientific literature, suggesting that adherence to a varied dietary model increases intestinal microbial diversity [43]. In numerous dietary models and, in particular, in the Mediterranean diet, microbial changes are already measured after three to four days of adherence to the diet [44].

From this study, it was concluded that among the various types of dried fruit, nuts are the ones that have the most significant effect on the microbiome due to their abundance of polyphenols and omega 3, considered prebiotics. The limitation of these studies is the numerical scarcity and the data collected in the microbiome after short-term intake of walnuts [41].

6. Conclusions

In this chapter, we wanted to highlight how the intake of dried fruit and, in particular, walnuts could be beneficial for our bodies.

We have evaluated numerous studies and articles, and we can summarize and highlight the following issues:

- Taking nuts with a controlled and daily dose as part of a balanced diet such as the Mediterranean diet helps prevent major metabolic and cardiovascular diseases

- The fibers in walnuts and dried fruit help fight excess gastric acidity and intestinal absorption of many nutrients and participate in the balance and growth of the intestinal microbial population. Dried fruit can be defined as "prebiotic," thanks to its fiber content.

- The microbiota metabolizes fibers and polyphenols derived from dried fruit, and thanks to these, it can improve the production of SCFA that participate in numerous physiological processes. Many bacterial families grow up expressing their beneficial effects of preventing and protecting countless metabolic diseases.

- Given the recommended daily rate, dried fruit can be a snack; as a high-calorie food, it can help combat asthenia, improve concentration during studies, be a deal-breaker for those on a low-calorie diet, and be an excellent snack for athletes to prepare for training or competition.

- The fats in dried fruit are omega 3 fats that protect our vessels and the heart. They are an excellent alternative for those who follow a vegan diet. In addition to fats, they are an excellent source of folic acid and mineral salts.

- The Mediterranean diet has always promoted and recommended the intake of dried fruit and fiber. It appears that dried fruit consumption is increasingly widespread and that the scientific community, hopefully, continues to study its beneficial effects.

Acknowledgements

The present work has been developed in collaboration with the VIRTUOUS project, funded by the European Union's Horizon 2020 research and innovation program under the Marie Sklodowska-Curie-RISE Grant Agreement No 872181 (https://www.virtuoush2020.com/).

Conflict of interest

The authors declare that the research was conducted without any commercial or financial relationships construed as a potential conflict of interest.

DOI: http://dx.doi.org/10.5772/intechopen.108567

Acronyms and abbreviations

SCFA	short-chain fatty acids
SFA	short fatty acids
IBS	irritable bowel syndrome
MUFA	monosaturated fatty acids
PUFA	polyunsaturated fatty acids
ALA	alpha linolenic acid
CVD	cardiovascular disease
EVO	extra virgin olive oil
EPA	eicosapentaenoic acid
LDL	low-density lipoprotein
HDL	high-density lipoprotein

Author details

Giorgio Calabrese[1]* and Cinzia Myriam Calabrese[2]

1 Department of Science and Technological Innovation, University of Eastern Piedmont, Alessandria, Italy

2 PolitoBIOMed Lab - Biomedical Engineering Lab, Politecnico di Torino, Italy

*Address all correspondence to: giorgiocalabrese@gcalabrese.it

References

[1] Nishida C. Diet, nutrition and the prevention of chronic diseases. World Health Organization Technical Report Series. 2003;**916**:i-viii 1-149. backcover. PMID: 12768890

[2] Leeming ER, Johnson AJ, Spector TD, Le Roy CI. Effect of diet on the gut microbiota: rethinking intervention duration. Nutrients. 2019;**11**(12):2862

[3] Putignani L, Del Chierico F, Petrucca A, Vernocchi P, Dallapiccola B. The human gut microbiota: a dynamic interplay with the host from birth to senescence settled during childhood. Pediatric Research. 2014;**76**(1):2-10

[4] Adak A, Khan MR. An insight into gut microbiota and its functionalities. Cellular and Molecular Life Sciences. 2019;**76**(3):473-493

[5] Jardine M. Nutrition considerations for microbiota health in diabetes. Diabetes Spectrum: A Publication of the American Diabetes Association. 2016;**29**(4):238-244

[6] Jandhyala SM, Talukdar R, Subramanyam C, Vuyyuru H, Sasikala M, Nageshwar RD. Role of the normal gut microbiota. World Journal of Gastroenterology. 2015;**21**(29):8787-8803

[7] Rowland I, Gibson G, Heinken A, et al. Gut microbiota functions: metabolism of nutrients and other food components. European Journal of Nutrition. 2018;**57**(1):1-24

[8] Davis-Richardson AG, Triplett EW. A model for the role of gut bacteria in the development of autoimmunity for type 1 diabetes. Diabetologia. 2015;**58**(7):1386-1393

[9] Vinolo MA, Rodrigues HG, Nachbar RT, Curi R. Regulation of inflammation by short chain fatty acids. Nutrients. 2011;**3**(10):858-876

[10] Luu M, Visekruna A. Short-chain fatty acids: bacterial messengers modulating the immunometabolism of T cells. European Journal of Immunology. 2019;**49**(6):842-848

[11] Vinolo MA, Rodrigues HG, Hatanaka E, Sato FT, Sampaio SC, Curi R. Suppressive effect of short-chain fatty acids on production of proinflammatory mediators by neutrophils. Journal of Nutritional Biochemistry. 2011;**22**(9):849-855

[12] Estruch R, Ros E, Salas-Salvadó J, Covas MI, Corella D, Arós F, et al. Primary prevention of cardiovascular disease with a Mediterranean diet. The New England Journal of Medicine. 2013;**368**(14):1279-1290. DOI: 10.1056/NEJMoa1200303 Epub 2013 Feb 25. Retraction in: N Engl J Med. 2018 Jun 21;378(25):2441-2442. Erratum in: N Engl J Med. 2014 Feb 27;370(9):886. Corrected and republished in: N Engl J Med. 2018 Jun 21;378(25):e34. PMID: 23432189.x

[13] Fraser GE, Sabaté J, Beeson WL, Strahan M. A possible protective effect of nut consumption on risk of coronary heart disease. Archives of Internal Medicine. 1992;**152**:1416-1424

[14] Kelly JH, Sabaté J. Nuts and coronary heart disease: an epidemiological perspective. British Journal of Nutrition. 2006;**96**:S61-S67

[15] Jiang R, Jacobs DR, Mayer-Davis E, et al. Nut and seed consumption and inflammatory markers in the multi-ethnic study of atherosclerosis. American Journal of Epidemiology. 2006;**163**:222-231

[16] http://data.europa.eu/eli/reg/2012/432/oj

[17] DeFelice SL. The nutraceutical revolution: Fueling a powerful, new international market. Como, Italy: Harvard University Advanced Management Program in Biomedical Research and Development; 1989

[18] Mitrou PN, Kipnis V, Thiébaut AC, et al. Mediterranean dietary pattern and prediction of all-cause mortality in a US population: results from the NIH-AARP Diet and Health Study. Archives of Internal Medicine. 2007;**167**:2461-2468

[19] Salas-Salvado J, Fernandez-Ballart J, Ros E, et al. A Mediterranean diet supplemented with nuts improves metabolic syndrome status: one year results of the PREDIMED randomized trial. Archives of Internal Medicine. 2008;**168**:2449-2458

[20] Nettleton JA, Steffen LM, Mayer-Davis EJ, Jenny NS, Herrington DM, Jacobs DR. Dietary patterns are associated with biochemical markers of inflammation and endothelial activation in the Multi-Ethnic Study of Atherosclerosis (MESA). American Journal of Clinical Nutrition. 2006;**6**:1369-1379

[21] Sabaté J, Ang Y. Nuts and health outcomes: new epidemiologic evidence. American Journal of Clinical Nutrition 2009 May;89(5):1643S-1648S. DOI: 10.3945/ajcn.2009.26736Q. Epub 2009 Mar 25. PMID: 19321572.

[22] Grosso G, Yang J, Marventano S, Micek A, Galvano F, Kales SN. Nut consumption on all-cause, cardiovascular, and cancer mortality risk: a systematic review and meta-analysis of epidemiologic studies. American Journal of Clinical Nutrition Apr 2015;**101**(4):783-793. DOI: 10.3945/ajcn.114.099515. Epub 2015 Feb 4. PMID: 25833976

[23] Bolling BW, Chen CY, McKay DL, Blumberg JB. Tree nut phytochemicals: composition, antioxidant capacity, bioactivity, impact factors. A systematic review of almonds, Brazils, cashews, hazelnuts, macadamias, pecans, pine nuts, pistachios and walnuts. Nutrition Research Reviews. 2011;**24**:244-275

[24] Brufau G, Boatella J, Rafecas M. Nuts: source of energy and macronutrients. British Journal of Nutrition. 2006;**96**(Suppl. 2):S24-S28

[25] Gornik HL, Creager MA. Arginine and endothelial and vascular health. Journal of Nutrition 2004;134 (Suppl. 10):2880S–7S; discussion 95S

[26] Ostlund RE Jr. Phytosterols in human nutrition. Annual Review of Nutrition. 2002;**22**:533-549

[27] Lu LW, Silvestre MP, Sequeira IR, Plank LD, Foster M, Middleditch N, et al. A higher-protein nut-based snack product suppresses glycaemia and decreases glycaemic response to co-ingested carbohydrate in an overweight prediabetic Asian Chinese cohort: the Tū Ora postprandial RCT. Journal of Nutritional Science. Apr 2021;**23**(10):e30. DOI: 10.1017/jns.2021.20. PMID: 34094511; PMCID: PMC8141680

[28] Luo C, Zhang Y, Ding Y, et al. Nut consumption and risk of type 2 diabetes, cardiovascular disease, and all-cause mortality: a systematic review and meta-analysis. American Journal of Clinical Nutrition. 2014;**100**:256-269

[29] Li S, Liu Y, Liu J, et al. Almond consumption improved glycemic control and lipid profiles in patients with type 2 diabetes mellitus. Metabolism. 2011;**60**:474-479

[30] Kendall C, Josse A, Esfahani A, et al. The impact of pistachio intake alone or

in combination with high-carbohydrate foods on post-prandial glycemia. European Journal of Clinical Nutrition. 2011;**65**:696

[31] Josse A, Kendall C, Augustin L, et al. Almonds and postprandial glycemia – a dose-response study. Metabolism. 2007;**56**:400-404

[32] Harvard Health Publishing: https://www.health.harvard.edu/staying-healthy/can-gut-bacteria-improve-your-health

[33] Liu H-N, Wu H, Chen Y-Z, Chen Y-J, Shen X-Z, Liu T-T. Altered molecular signature of intestinal microbiota in irritable bowel syndrome patients compared with healthy controls: a systematic review and meta-analysis. Digestive and Liver Disease. 2017;**49**:331-337. DOI: 10.1016/j.dld.2017.01.142

[34] Attaluri A, Jackson M, Valestin J, Rao SSC. Methanogenic flora is associated with altered colonic transit but not stool characteristics in constipation without IBS. American Journal of Gastroenterology. 2010;**105**:1407-1411. DOI: 10.1038/ajg.2009.655

[35] Jalanka-Tuovinen J, Salonen A, Nikkilä J, Immonen O, Kekkonen R, Lahti L, et al. Intestinal microbiota in healthy adults: Temporal analysis reveals individual and common core and relation to intestinal symptoms. PLoS One. 2011;**6**:e23035. DOI: 10.1371/journal.pone.0023035

[36] Wong JMW, de Souza R, Kendall CWC, Emam A, Jenkins DJA. Colonic health: fermentation and short chain fatty acids. Journal of Clinical Gastroenterology. 2006;**40**:235-243. DOI: 10.1097/00004836-200603000-00015

[37] Lamuel-Raventos RM, Onge MS. Prebiotic nut compounds and human microbiota. Critical Reviews in Food Science and Nutrition. 2017;**57**(14):3154-3163. DOI: 10.1080/10408398.2015.1096763. PMID: 27224877; PMCID: PMC5646185

[38] Creedon AC, Hung ES, Berry SE, Whelan K. Nuts and their effect on gut microbiota, gut function and symptoms in adults: a systematic review and meta-analysis of randomised controlled trials. Nutrients. 2020;**12**(8):2347. DOI: 10.3390/nu12082347. PMID: 32781516; PMCID: PMC7468923

[39] Holscher HD, Guetterman HM, Swanson KS, An R, Matthan NR, Lichtenstein AH, et al. Walnut consumption alters the gastrointestinal microbiota, microbially derived secondary bile acids, and health markers in healthy adults: a randomized controlled trial. Journal of Nutrition. 2018;**148**:861-867. DOI: 10.1093/jn/nxy004

[40] Tindall AM, McLimans CJ, Petersen KS, Kris-Etherton PM, Lamendella R. Walnuts and vegetable oils containing oleic acid differentially affect the gut microbiota and associations with cardiovascular risk factors: follow-up of a randomized, controlled, feeding trial in adults at risk for cardiovascular disease. Journal of Nutrition. 2020;**150**:806-817. DOI: 10.1093/jn/nxz289

[41] Fitzgerald E, Lambert K, Stanford J, Neale EP. The effect of nut consumption (tree nuts and peanuts) on the gut microbiota of humans: a systematic review. British Journal of Nutrition 2021;**125**(5):508-520. DOI: 10.1017/S0007114520002925. Epub 2020 Jul 27. PMID: 32713355

[42] Yatsenko T, Rey FE, Manary MJ, et al. Human gut microbiome viewed across age and geography. Nature. 2012;**486**:222-227

[43] Garcia-Mantrana I, Selma-Royo M, Alcantara C, et al. Shifts on gut microbiota associated to Mediterranean diet adherence and specific dietary intakes on general adult population. Frontiers in Microbiology. 2018;**9**:890

[44] Walker AW, Ince J, Duncan SH, et al. Dominant and diet-responsive groups of bacteria within the human colonic microbiota. sISME Journal. 2011;**5**:220-230

Chapter 7

Therapeutic Potential of Chocolate Coated Chinese Chestnut (A Nut Cum Fruit)

Sunita Mishra, Gunjan Gupta and M.A. Firdaus

Abstract

Castanea mollissima, usually known as Chinese chestnut tree, is a medium-tall tree with glossy, dark green leaves. The tree produces edible and delicious nuts called chestnuts or Chinese chestnuts. Since many years, China has loomed as one of the largest bowls of Chinese chestnut in the world. Although the consumption of chestnut has increased in India too in the states like Uttarakhand which shares border with China. Chestnuts are a rich source of health-promoting nutrients including vitamin A, vitamin E, calcium, magnesium, iron, zinc, copper, manganese, etc. They got dark browny shells and shiny spiral husk which is to be broken before eating. Castanea mollisima aids in the development of muscle and the prevention of dental issues like pyorrhea. The leafy portions of the nut are used to treat fevers, relieve persistent coughs and respiratory conditions, and protect blood vessels and capillaries from injury and swelling. In order to analyse the nutritional content, organoleptic qualities, and microbiological quality of Chinese chestnut after coating with dark chocolate, this study will examine the therapeutic potential of Chinese chestnut, making it an innovative and great choice to consume for people of all ages.

Keywords: Chinese chestnut, nutritional value, organoleptic properties, microbiological quality, therapeutic potential

1. Introduction

Chestnuts are members of the Fagaceae family. The majority of chestnut trees are found in North America, South Europe, and Easter and Southwest Asia. Among all species of chestnut, the yearly production of the Chinese chestnut (Castanea mollissima Blume) is around 925,000 t, as compared to 108,000 t for the European chestnut (C. sativa Miller) and 55,800 t for the North American and South American chestnut (C. dentata Borkh) [1–3]. The Yan Mountain region in Northern China has the low temperatures and altitudes above 500 m that are ideal for Chinese chestnut production. Chestnuts are a great source of lot of nutritional content. According to Yang et al. [4], fresh Chinese chestnut fruits contain 52.0% water, 42.2% carbohydrates, 4.2% proteins, and 0.7% lipids, whereas Spanish chestnuts contain 42.2–59.8% starch, 9.5–22.2% total sugar,

Figure 1.
Chinese chestnuts.

4.8–6.9% crude protein, and 1.7–40% dry matter, depending on the cultivar and region [5]. According to the ancient encyclopedia, China Compendium of Materia Medica (Ben Cao Gang Mu) from the Ming Dynasty, Chinese chestnut fruits aid in enhancing kidney function (A.D. 1590). According to numerous health studies, chestnut fruits and possibly other extracts from chestnut trees offer a lot of potential as functional foods or food additives [1, 3]. The Chinese Chestnut (**Figure 1**), also scientifically called as C. mollissima (mollissima is Latin word for "soft"), is primarily grown in East Asia but is also indigenous to places like Taiwan and Korea. The abundant pubescence on the abaxial sides of its leaves and buds is how it got its name.

This study aims to evaluate the nutritional content, microbiological quality, and organoleptic qualities of dark chocolate-coated chestnut, making it a novel food option for consumers of all ages.

1.1 Pharmacological activities of Chinese chestnuts

Chestnut is essential and useful in the treatment of a number of ailments (**Figure 2**). Therefore the pharmacological activity is presented here based on the percentage of diseased states [9, 10].

1.2 Effect of cooking on Chinese chestnuts

The nutritional content of six cultivars of Cannabis sativa was thoroughly measured, including starch, energy, fat & fibre [11], crude protein, amino acids, phenolic phytochemicals [12], minerals, sugars, antioxidant carotenoids and other vitamins [1, 3] after cooking during four major stages of industrial processing (fresh, after storing for three months at ±0°C and maintaining the relative humidity at 90%, after industrial peeling through flame or fire at 800–1000°C, and after freezing in a tunnel with CO^2 flow at −65°C [1, 3].

According to a study [13], the vitamin C content of fresh chestnuts ranged from 400 to 693 mg/kg dry weight for several European cultivars. The boiling and roasting processes resulted in significant reductions of 25–54% and 2–77%, respectively. A typical traditional way for preparing Chinese chestnuts is to either fry them shelled with other ingredients (tang chao li zi) or unshelled with sugar. In East Asia, a sort of

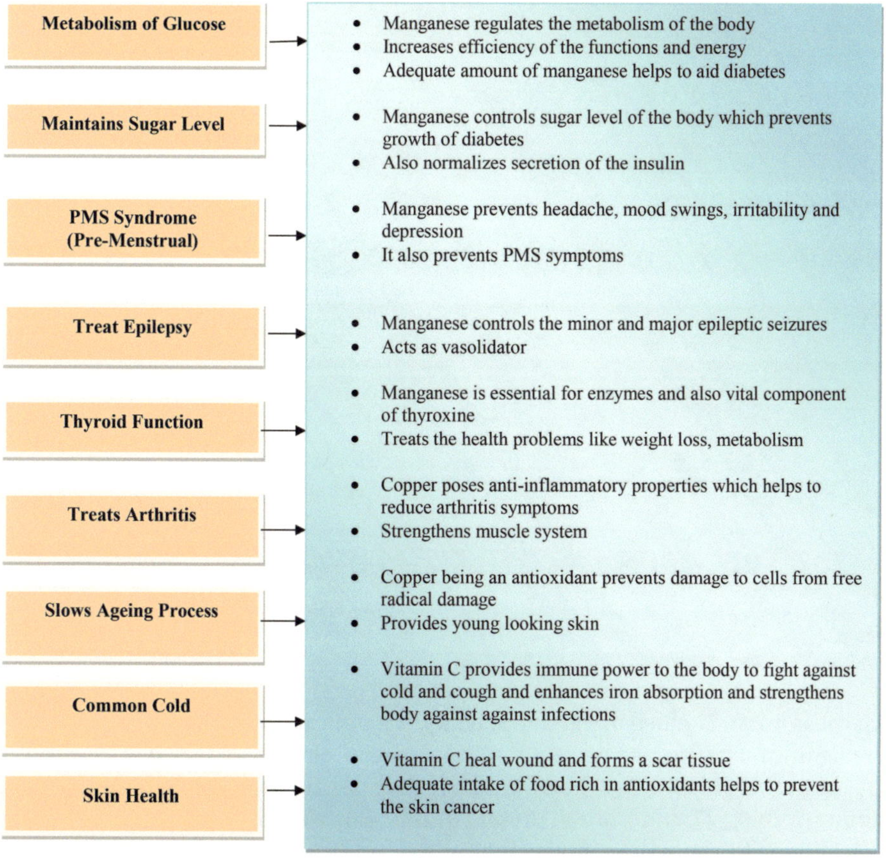

Figure 2.
Pharmacological activities of Chinese chestnuts [6–8].

chestnut kernel that has been processed industrially and packaged is currently very popular and is typically cooked by boiling.

Even though a fresh chestnut's kernel does not smell strongly, after being thermally processed, it has a strong flavour. The flavour extract of boiled and roasted Chinese chestnuts, respectively, yielded a total of 30 to 33 components, including hydrocarbons, ketones, alcohols, aldehydes, furans, pyranone, and acids [14]. Important aroma impact components from roasted Italian chestnuts include monoterpenes and derivatives of butane, pentane, hexane, and heptane [15].

1.3 Dark chocolate

One of the roasted and ground cocoa products is chocolate. It is manufactured in block, paste, or liquid form and added to products to improve the flavour of that product. It has a lot of calories and a healthy amount of fat. However, dark chocolate, one form of chocolate, is thought to be highly healthy to eat. Dark or black chocolates can be eaten as it is, or used in recipes, for which somewhat thicker-packed bars are sold in market. The seeds of the tropical tree Theobroma cacao are used to make raw or processed foods, including dark chocolate, which has been a popular food for centuries. Depending on how

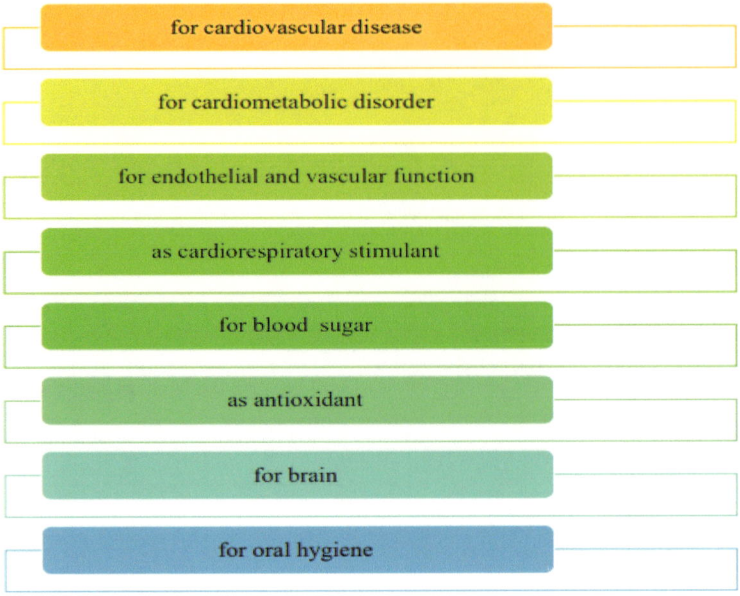

Figure 3.
Health benefits of dark chocolate.

much cocoa butter is added to the components, it is typically bitter or semisweet. Some of the significant health advantages of eating dark chocolate are depicted in **Figure 3**. Many healthful components that are good for the body are found in dark chocolate. Vitamins A, B1, C, D, and E are all present in abundance in compounds found in chocolate. Additionally, chocolate contains the antioxidants phenol and flavonoids, as well as minerals like calcium, potassium, iron, omega 3 and 6, and high magnesium, which aid to somewhat lessen premenstrual symptoms and menstrual pain [16].

1.4 Flaxseeds

Flaxseed, whose scientific name is Linum usitatissimum, which means "useful," is also known as Alsi or Jawas in India and has a crunchy texture and nutty flavour. It has been grown for its fibre content, other nutritional benefits, and therapeutic applications. **Figure 4** illustrates the advantages of flaxseeds for health. As the building block for other compounds that help in the prevention of inflammation, ALA (alpha-linolenic acid) can also help protect the blood vessels against inflammatory damages. Numerous studies have demonstrated that dietary flaxseeds can raise blood levels of ALA, even when they are added to baked goods like breads, muffins, or other recipes. Two more omega 3 fatty acids, eicosapentaenoic acid (EPA) and docosapentaenoic acid, have also been shown to increase in the blood when flaxseeds are consumed (DPA). Increased blood levels of EPA and DPA help in the inflammatory protection that arises in many degenerative illnesses [17–23].

1.5 Objective

- To investigate the confection's nutritional value.
- To evaluate the confection's microbiological quality.

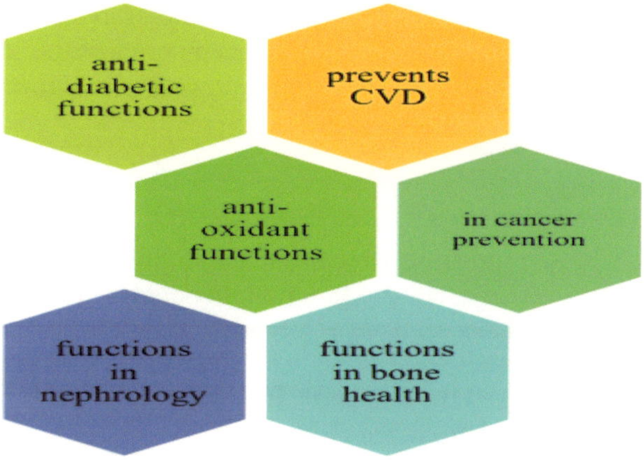

Figure 4.
Health benefits of flaxseeds.

- To develop and standardise the Chinese chestnut product.
- To examine the sensory acceptability of the product and its packaging.

2. Materials and methods

2.1 Sample preparation

Fresh Chinese Chestnuts were collected from the local market of Nainital, Uttarakhand, India. The chestnuts were first washed to eliminate any dust or fine particles. They were then sun dried for 3–4 days at 32°C. The ingredient's shelf life was extended by drying as there was no water activity and no moisture present. The flaxseeds were purchased from the local store in the city of Bazpur, Uttarakhand, India. After dry roasted, they were later ground into powder. The chocolate compound bar was bought from the local market of Lucknow city of Uttar Pradesh, India.

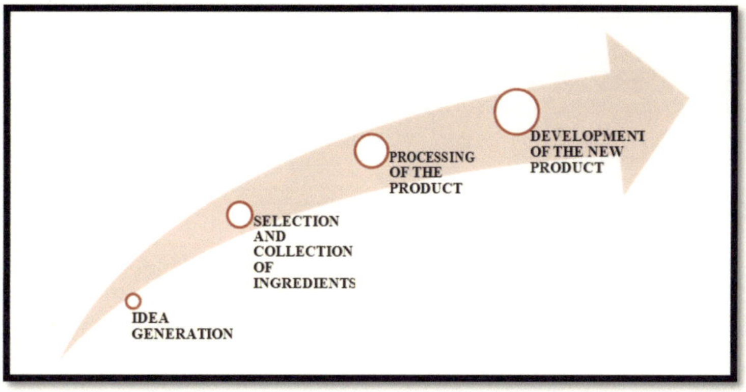

Figure 5.
Preparation of mix.

The chocolate compound was tempered at 42°C for further use in the process of product development. To eliminate variations in their composition and avoid quality differences, the procurement was completed in a single lot. In **Figure 5**, the stages of product development are shown.

2.2 Preparation of mix

The sun-dried Chinese chestnuts were roasted after being ground into a grainy powder. The flaxseeds were roasted and ground into powder in a same manner. Externally added sugar and any kind of preservatives were avoided. The dried and roasted chestnuts and flaxseeds were combined in a ratio of 80:20 with the roasted powders. **Figure 6** summarises the schematic flow diagram of the formulation and preparation of the chocolate coated chestnut.

2.3 Nutritional evaluation

The prepared confection was first ground and blended. The sample was then analysed using several AOAC approved methods (1997 & 2000) of analysis to determine the presence of different nutrients.

2.3.1 Moisture

By drying a sample in oven at 105°C until a constant weight was detected, the moisture content was examined using the following formula:

Moisture% = (initial weight-final weight x 100) /weight of the sample.

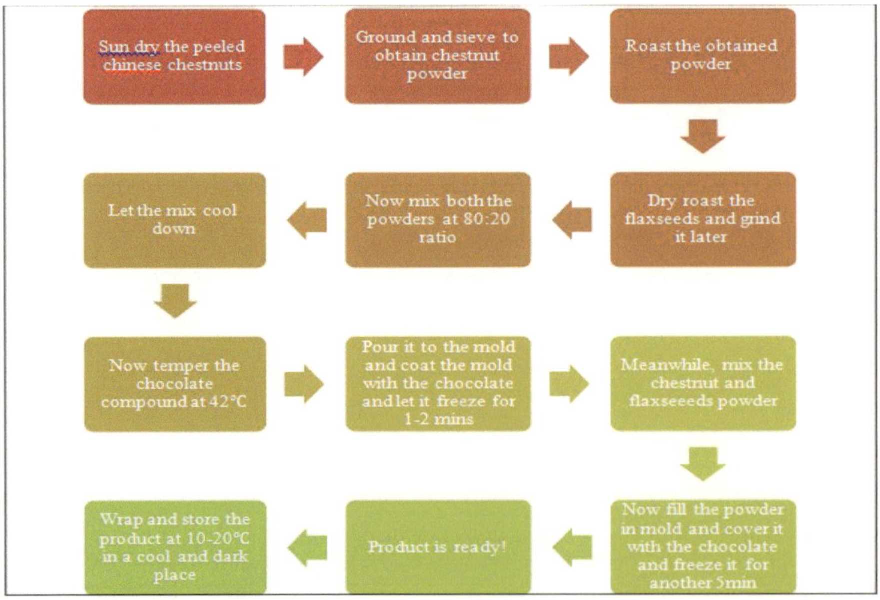

Figure 6.
Sample preparation.

2.3.2 Protein

The total nitrogen content was ascertained using the Kjeldahl method. The nitrogen percentage was first determined, and then it was multiplied by a factor of 4.86, or 4.86 x N to determine the protein content percentage (nitrogen percentage).

2.3.3 Carbohydrate

The content of the readily available carbohydrate was calculated by differences i.e., by subtracting from 100 the total of the values for moisture, protein, ash and fat (per 100gm).

2.3.4 Fat

As a crude ether extract of the dry material, fat is extracted from the sample. Firstly, the dry sample weighs about 5–10 gm into a thimble and covered with cotton. The thimble is then put in a Soxhlet Apparatus and extracted for 15 to 16 hours with anhydrous ether. After filtering the ether extract into the weighted conical flask, it was cleaned 4–5 times. After the ether has been eliminated through evaporation, the remaining material is dried in an oven at 80 to 100°C, cooled in a desiccator, and then weighed.

Fat content (g/100gm of sample) = (weight of ether extract x 100) /Weight of sample (equivalent to fresh sample taken).

2.3.5 Ash

Ash content was determined gravimetrically of the sample residue after ignition in vacuum oven at 512°C to constant weight.

Ash content (g/100 gm sample) = (Weight. of the ash x 100) /Weight. of the sample taken.

2.3.6 XRD (X-ray diffraction)

A laboratory-based method known as X-Ray Diffraction (XRD) is typically used to analyse unit cell dimensions and identify crystalline materials [24]. It is frequently used in a variety of disciplines, including engineering, geology, material science, and environmental sciences, to quickly identify unidentified crystalline substances using Bruker D8-eco Advance XRD (usually within 20 minutes).

It follows *Bragg's Law:* $n\lambda = 2d \sin\theta$ formula.

It is necessary to use pure, finely ground, and homogenised samples to determine the bulk composition. This method can also be used for the identification of fine-grained minerals, quantitative determination of modal quantities of minerals in a sample, comprehensive characterisation of crystalline materials, and identification of unit cell dimensions.

2.3.7 Chocolate bloom

When chocolate blooms, it means that the chocolate was not stored properly. However, one can also eat the bloomed chocolates. The two types of blooms that were seen in this study were:

a. Fat, which appears as grey-white blotches and streaks on chocolate when it is exposed to heat.

b. Sugar makes chocolate feel scratchy and occurs when chocolate is stored in humid conditions.

2.4 Microbiological analysis

The dark-chocolate chestnut shells (DCC) which were stored at 7°C and 25 ± 2°C examined for the microbiological analysis. The reference sample (REF) was the marketed fruit and nut dark chocolate that was purchased from a nearby market. Samples were analysed for TPC (Total Plate Count), Salmonella spp., Escherichia coli, as well as yeast and moulds. The growth media used for the micro- organisms identification were: Nutritive Agar (for Total Plate Count), MacConkey Agar (for E. coli), Czapek Agar (for yeasts and moulds) and Salmonella-Shigella Agar, Baird-parker Agar (for Staphylococcus aureus).

For each of the experiments mentioned, about 3 g of sample was powdered and homogenised in 30 ml of broth in a sterile flask before being incubated for 24 hours at 37°C. All the samples were aseptically removed and progressive dilution of 102 decimal were prepared and later plated on appropriate media. The plates were then incubated for a further 48 hours at 37°C, during which time the colonies were manually counted using a permanent marker pen. It was shown that there were variations between the two samples at a 90% significance level [25].

2.5 Organoleptic properties

The distribution of the samples was done as follows for this purpose:

- T1 Drinking Chocolate Chestnut Shells
- T2 Chestnut Shells from Hershey
- T3 Chestnut Shells from Dark Chocolate

2.5.1 Preparation of the 'Drinking Chestnut Chocolate Shells'

Ingredients included: 250gm of ground Chinese chestnut, 50gm of roasted and ground flaxseeds, 2 tablespoons of ghee to roast the ingredients and give the chocolate a glossy appearance, 50gm of drinking chocolate powder, and 150gm of chocolate compound. No sugar was added before, during, or after preparation. **Figure 6** gives the step-by-step procedure for the preparation of this product. The final product was covered in glossy RED wrapping foils with the code T1 (SAMPLE-1) (**Figures** 7 and **8**).

2.5.2 Preparation of the 'Hershey's Chestnut Chocolate Shells'

The ingredients included: 250 gm of ground Chinese chestnut, 50 gm of roasted and ground flaxseeds, 2 tablespoons of ghee to roast the ingredients and give the chocolate a glossy appearance, 50 gm of Hershey Chocolate Syrup, and 150 gm of chocolate compound. No sugar was added before, during or after preparation. **Figure 6** gives the step-by-step procedure for the preparation of this product. The final product was covered in glossy YELLOW wrapping foils with the code T2 (SAMPLE-2) (**Figures 8** and **9**).

Figure 7.
Drinking chestnut chocolate shells.

Figure 8.
Wrapped chocolates: T1 – Red foil, T2 – Yellow foil, T3 – Blue foil.

2.5.3 Preparation of the 'Dark Chestnut Chocolate Shells'

The ingredients included: 250 gm of ground Chinese chestnut, 50 gm of roasted and ground flaxseeds, 2 tablespoons of ghee to roast the ingredients and give the chocolate a glossy appearance, 50 gm of Cocoa Powder, and 150 gm of Dark Chocolate compound. No sugar was added before, during or after preparation. **Figure 6** gives the step-by-step procedure for the preparation of this product. The final product was covered in glossy BLUE wrapping foils with the code T3 (SAMPLE-3) (**Figures 8** and **10**).

Figure 9.
Hershey's chestnut chocolate shells.

Figure 10.
Dark chestnut chocolate shells.

2.5.4 Packaging

Food packaging, one of the most accurate methods, helps in food containment as well as protection against several physical, chemical, and biological hazards to the food. Today, it is our duty as responsible and educated consumers to determine whether the product we spend our money on is clean, appropriately packaged, and sealed before being used.

2.5.5 Tools

Cardboard secondary packaging boxes and colourful aluminium foils.

2.5.6 Procedure

The chocolates were wrapped using the primary packaging material which were colourful foils assigned according to their sample codes, such as T1 for red foil,

T2 for yellow foil, and T3 for blue foil with proper hygiene (**Figure 8**). After that, it was placed in the secondary packaging material (cardboard boxes) and were weighed 100gm each per boxes. The sensory evaluation for this study was done in the Department of Food and Nutrition of BBAU Lucknow, by the efficiently trained faculty members and the technique used was Hedonic Scale [26].

3. Result and discussion

3.1 Nutritional composition

Table 1 shows the nutritional composition of dark chocolate-coated Chinese chestnut. Except for the moisture content, all compositions were measured using dry materials (**Table 1**). The prepared chocolate contained no extra ingredients when compared to commercial fruit and nut dark chocolate, which is made up of preservatives and other sweeteners. The dark chocolate-coated chinese chestnut is therefore highly fascinating and appears to be beneficial for human health, as proved by its overall high nutritional quality.

By using a glancing angle X-ray diffractometer with monochromatic Cu-K as the radiation source (45 kV and 20 mA), the XRD pattern of the prepared sample was recorded (**Figure 11**). As shown in **Figure 11**, the XRD measurements were made in the range of 2θ from 5° to 90°.

The sample's obtained diffraction pattern reveals its amorphous nature. Using Debye Scherrer's formula, the sample's crystallite size (Dhkl) was estimated-.

$$D_{hk1} = \frac{0.9\lambda}{\beta_{hkl}{}^{Cos}(\theta_{hkl})}$$

Here λ represents the X-ray wavelength, θhkl indicates the Bragg diffraction angle and βhkl represents the full width at half maximum (FWHM) of the major peaks in XRD pattern in radians.

The sample's crystallite size was determined to be 76.30 nm.

3.2 Chocolate bloom

The chocolate bloom was shown for both of its components—fat and sugar—at a temperature of 26°C, which is a good storage temperature. The blooming was also avoided by keeping the storage area dry.

Carbohydrate	**53.55 gm**
Energy	449 Kcal
Protein	8.3 gm
Total fat	22.4 gm
Ash	5.69%
Moisture	3.6%

Table 1.
Nutritional composition of the product.

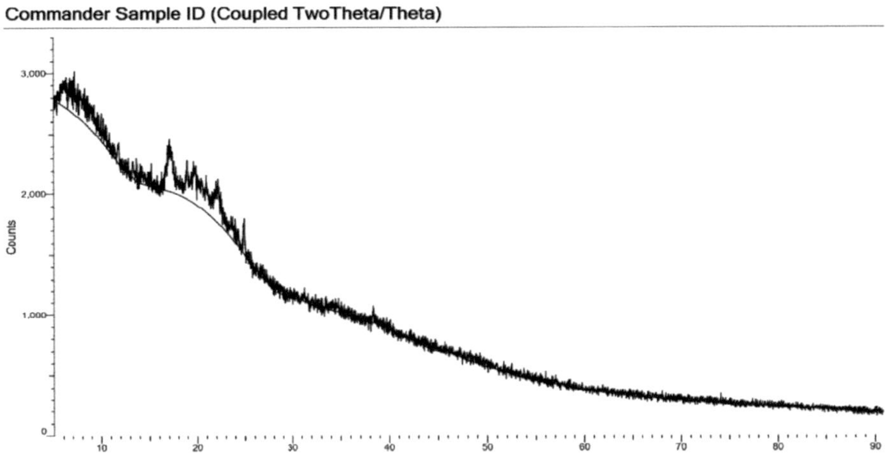

Figure 11.
X-ray diffraction for dark chocolate coated Chinese chestnut.

3.3 Microbiological quality

The microbiological analysis of the dark chocolate coated Chinese chestnut and fruit & nut dark chocolate both were analysed under controlled circumstances at 7°C and 25 ± 2°C respectively. The table in **Figure 12** provided the results. Yeasts and moulds were not present, and the TPC was lower than that of the fruit and nut dark chocolate used as the reference sample for the testing (**Figure 12**). Additionally, the sample did not include any colonies of the bacteria S. aureus, Salmonella, or *E. coli*. The lack of yeast, moulds, and TPC confirmed that the raw material was treated carefully at every stage, from collection till processing and analysis of the developed product.

3.4 Organoleptic properties

Expert panel members conducted a sensory assessment of processed and packaged Chinese chestnut shells using a hedonic scale and marking on the following four criteria:

3.4.1 Body and texture

According to **Figure 13**, T3, which received the highest scores for body and texture, was the sample that the expert panellists found to be the most acceptable. Next, sample T1 and T2, respectively.

	Conditions	Storage duration	TPC (aerobic) (CFU/g)	*E. coli* (CFU/g)	*Salmonella* spp.	Yeasts and molds (CFU/g)	*Stapphylococcus aureus*	*Shigella* spp.
DCC	7°C; (82±5)% RH	Day 15	$1.2x\ 10^2$	<20	ND	<20	ND	ND
	25°C±2°C; (25±2)% RH	Day 30	$1.2x\ 10^2$	<30	ND	<20	ND	ND
REF	7°C; (82±5)% RH	Day 15	$1.2x\ 10^2$	<20	ND	<20	ND	ND
	25°C±2°C; (25±2)% RH	Day 30	$1.2x\ 10^2$	<30	ND	<20	ND	ND

Figure 12.
Microbiological evaluation of chocolate.

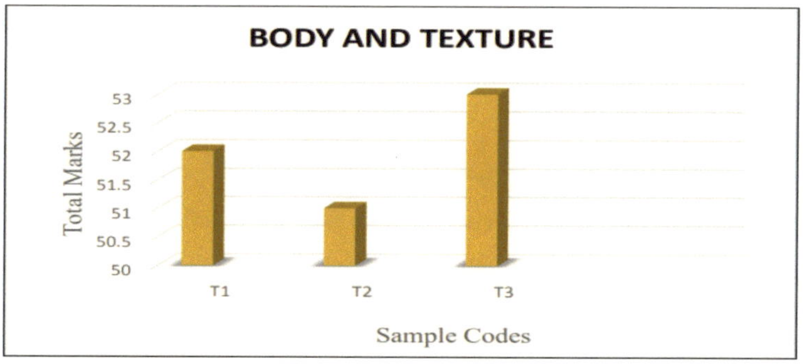

Figure 13.
Graphical representation of body and texture.

3.4.2 Colour and appearance

Figure 14 demonstrates that T3, which received the highest scores for colour and appearance, was the sample that the expert panel members found to be the most acceptable. Samples T2 and T1 are then, respectively.

3.4.3 Flavour and aroma

Figure 15 demonstrates that T3, which received the highest ratings for flavour and aroma, was the sample that the expert panellists found to be the most acceptable. Samples T1 and T2 combined were next.

3.4.4 Overall acceptability

Figure 16 demonstrates that the sample that received the best rating for overall acceptability from the expert panel members was T3. Samples T1 and T2 are then, respectively.

3.4.4.1 Overall calculation

By using the sensory evaluation scores provided by each panellist on the hedonic scale, an overall calculation was essentially made to determine the acceptability of the

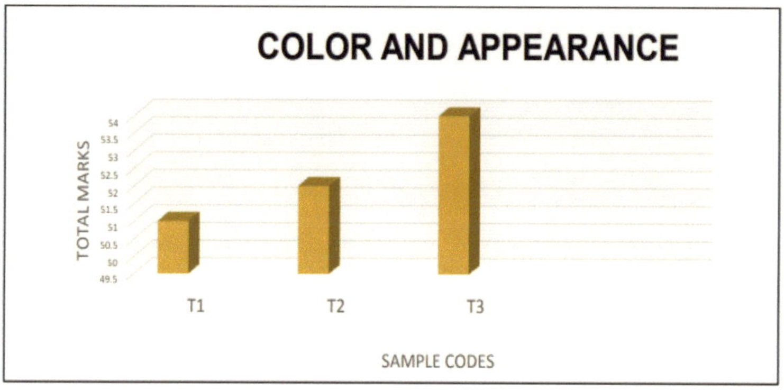

Figure 14.
Graphical representation of colour and appearance.

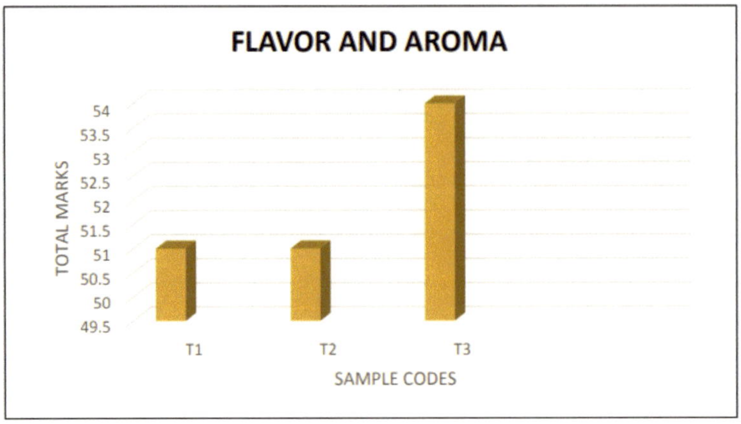

Figure 15.
Graphical representation of flavour and aroma.

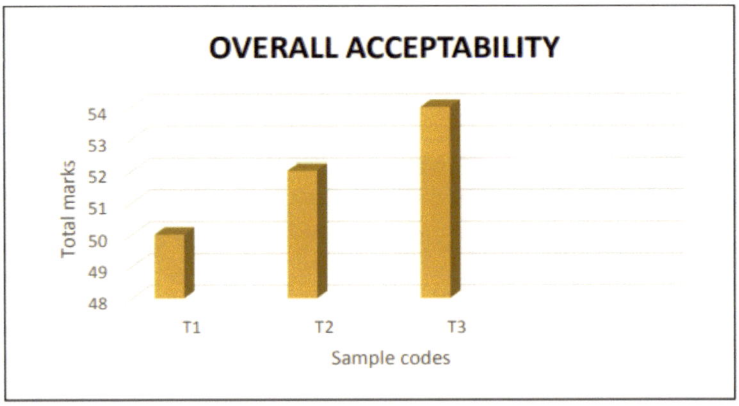

Figure 16.
Graphical representation of overall acceptability.

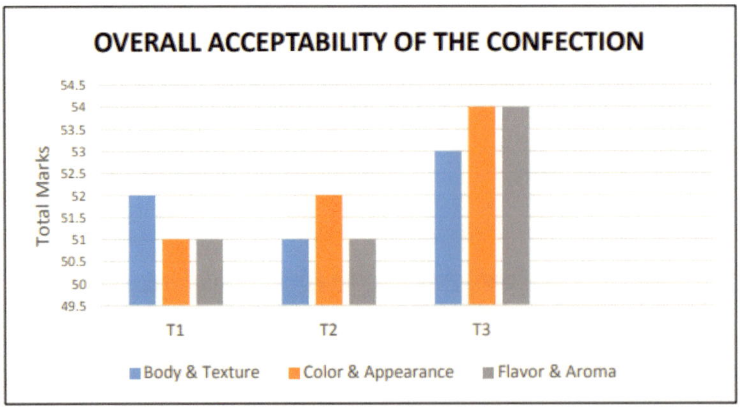

Figure 17.
Graphical representation of hedonic scale rating for overall acceptability of the confection for all the criterion.

final product in all quality terms. The scoring of texture, aroma, colour, flavour, and taste were calculated in the table, letting us to perform statistical analysis and obtain standard deviation, average, and other calculations (**Figure 17**).

4. Summary and conclusion

The dark chocolate coated Chinese chestnut known as dark chocolate Chinese chestnut shells has been prepared successfully and the result of its proximate analysis gives enough nutritional values to be considered as a novel and interesting product that should be introduced among the people. Comparing this study to earlier ones reveals that, in addition to roasting and boiling, etc., Chinese chestnuts can also be consumed in a variety of ways in India. According to microbiological studies, chocolate can help extend a product's shelf life and can be kept for up to 5 months without the need of any additional preservatives when kept in the right storage circumstances. Chinese chestnut being rich in nutrition is safe to consume and recommended for arthritis and many more health conditions. Furthermore, flaxseed is a wonderful source of fibre and numerous other nutritive elements in addition to being a cure for conditions like cardiovascular disease. Both when combined together can useful and a healthy option for people [27–32]. As a result, the manufactured confection, Chocolate Chessy Balls, is a way to spread awareness of the nut and its health benefits among people. The purpose behind the development of this product was both to introduce consumers to an exotic nut and to inform them about the benefits of the Chinese Chestnut. Organoleptical indicator is an useful way to ensure quality by assessing the product's appearance, flavour, and acceptability. All of the panellists approved the chocolates and not only expressed their appreciation for them but also made suggestions for additional recommendations to raise awareness of the miraculous nut native to country like China, Japan, Korea to the people of India's states like Uttar Pradesh.

Authors' contributions

Author[1] done all the testing. Author[2] analysed the data. Authors[3] wrote the main manuscript. All authors read and approved the final manuscript.

Funding

This research received no external funding.

Declaration of competing interest

The authors declare that there is no conflict of interests among them regarding the publication of this paper.

Author details

Sunita Mishra*, Gunjan Gupta and M.A. Firdaus
Department of Food and Nutrition, School of Home Science, Babasaheb Bhimrao Ambedkar (A Central University), Lucknow, India

*Address all correspondence to: sunitabbau@gmail.com

References

[1] De Vasconcelos MCBM, Nunes F, Viguera CG, Bennett RN, Rosa EAS, Ferreira-Cardoso JV. Industrial processing effects on chestnut fruits (Castanea sativa mill.) 3. Minerals, free sugars, carotenoids and antioxidant vitamins. International Journal of Food Science & Technology. 2010b;**45**(3):496-505

[2] Fernandes Â, Barreira JCM, Antonio AL, Bento A, Botelho ML, Ferreira ICFR. Assessing the effects of gamma irradiation and storage time in energetic value and in major individual nutrients of chestnuts. Food and Chemical Toxicology. 2011;**49**(9):2429-2432

[3] De Vasconcelos MCBM, Bennett RN, Rosa EAS, Ferreira-Cardoso JV. Composition of European chestnut (Castanea sativa mill.) and association with health effects: Fresh and processed products. Journal of the Science of Food and Agriculture. 2010a;**90**(10):1578-1589

[4] Yang YX, Pan XC, Wang GY. China Food Composition. 2nd ed. Beijing: Peking University Medical Press; 2009. p. 80

[5] Pereira-Lorenzo S, Ramos-Cabrer AM, Díaz-Hernández MB, Ciordia-Ara M, Ríos-Mesa D. Chemical composition of chestnut cultivars from Spain. Scientia Horticulturae. 2006;**107**(3):306-314

[6] Vishal P, Shivendra Kumar D, Yusuf AJ. Chocolates as dosage form-an overview. International Journal of Pharmaceutical and Research Sciences. 2012;**1**(6):397-412

[7] SatoA,TruongVD,JohanningsmeierSD, Reynolds R, Pecota KV, Yencho GC. Chemical constituents of sweet potato genotypes in relation to textural characteristics of processed French fries. Journal of Food Science. 2017;**83**(1):60-73

[8] Shahkhalili Y, Duruz E, Acheson K. Digestibility of cocoa butter from chocolate in humans. European Journal of Clinical Nutrition. 2000;**54**(2):120-125

[9] Barreira JCM, Casal S, Ferreira ICFR, Oliveira MBPP, And Pereira JA. Nutritional, fatty acid and triacylglycerol profiles of Castanea Sativa mill. Cultivars: A compositional and Chemometric approach. J. Agr. Food Chem. 20096;57:2836-2842

[10] Baron AM, Donnerstein RL, Samson RA, Baron JA, Padnick JN, Goldberg SJ. Hemodynamic and electrophysiologic effects of acute chocolate ingestion in young adults. American Journal of Cardiology. 1999;**84**(3):370-373

[11] De Vasconcelos MCBM, Bennett RN, Rosa EAS, Ferreira-Cardoso JV. Industrial processing effects on chestnut fruits (Castanea sativa mill.). 1. Starch, fat, energy and fibre. International Journal of Food Science & Technology. 2009a;**44**(12):2606-2612

[12] De Vasconcelos MCBM, Bennett RN, Rosa EAS, Ferreira-Cardoso JV. Industrial processing effects on chestnut fruits (Castanea sativa mill.). 2. Crude protein, free amino acids and phenolic phytochemicals. International Journal of Food Science & Technology. 2009b;**44**(12):2613-2619

[13] Barros AIRNA, Nunes FM, Gonçalves B, Bennett RN, Silva AP. Effect of cooking on total vitamin C contents and antioxidant activity of sweet chestnuts (Castanea sativa mill.). Food Chemistry. 2011;**128**(1):165-172

[14] Morini G, Maga JA. Volatile compounds in roasted and boiled

Chinese chestnuts (Castanea mollissima). Lebens-mittel-Wissenschaft und-Technologie. 1995;**28**(6):638-640

[15] Krist S, Unterweger H, Bandion F, Buchbauer G. Volatile compound analysis of SPME headspace and extract samples from roasted Italian chestnuts (Castanea sativa mill.) using GC-MS. European Food Research and Technology. 2004;**219**(5):470-473

[16] Arjanggi E, Tih F, Suwindere W. Black chocolate consumption reduces subjective symptoms in 18-22 years old females with premenstrual syndrome. Journal of Medicine & Health. 2015;**1**:76-84. DOI: 10.28932/jmh.v1i1.502

[17] Dodin S, Cunnane SC, Mâsse B, Lemay A, Jacques H, Asselin G, et al. Flaxseed on cardiovascular disease markers in healthy menopausal women: A randomized, double-blind, placebo-controlled trial. Nutrition. 2008 Jan;**24**(1):23-30. DOI: 10.1016/j.nut.2007.09.003

[18] AOAC. Official Methods of Analysis. Arlington VA, USA: Association of Official Analytical Chemists; 1997

[19] AOAC. Official Methods of Analysis. Arlington VA, USA: Association of Official Analytical Chemists; 2000

[20] Riediger ND, Othman R, Fitz E, Pierce GN, Suh M, Moghadasian MH. Low n6:n3 fatty acid ratio, with fish or flaxseed oil, in high fat diet improves plasma lipids and beneficially alters tissue fatty acid composition in mice. European Journal of Nutrition. 2009;**47**:153-160

[21] Morris DH. Flax—A Health and Nutrition Primer. 4th ed. 2007. Flax Council Of Canada. ISBN: 0969607350

[22] Rubilar M, Gutiérrez C, Verdugo M, Shene C, Sineiro J. Flaxseed as a source of functional ingredients. Journal of Soil Science and Plant Nutrition. 2010;**10**:373-377

[23] Deeksha K, Mishra S. Utilization of mango and its by-products by different processing methods. Asian Journal of Science and Technology. 2018;**09**(10):8896-8901

[24] Bhaskar R, Li J, Lijun X. A comparative study of particle size dependency of IR and XRD methods for quartz analysis. American Industrial Hygiene Association Journal. 1994;**55**(7):605-609. DOI: 10.1080/15428119491018682

[25] Gupta G, Mishra S. Nutritional and microbiological evaluation of dark chocolate coated Chinese chestnut. Journal of Basic and Applied Research International. 2021;**27**(5):29-37

[26] Gupta G, Mishra S, Chakrobarty A, Dubey P, Shankar P. Organoleptic properties of chocolate coated Chinese chestnut. Current Journal of Applied Science and Technology. 2021;*40*(14):54-63. DOI: 10.9734/cjast/2021/v40i1431403

[27] Adlercreutz H. Lignans and human health. Critical Reviews in Clinical Laboratory Sciences. 2007;**44**(5-6):483-525

[28] Bruinsma K, Taren DL. Chocolate: Food or drug? Journal of the American Dietetic Association. 1999;**99**(10):1249-1256. DOI: 10.1016/S0002-8223(99)00307-7

[29] Dribnenki JCP, McEachern SF, Chen Y, Green AG, Rashid KY. 2149 Solin (low linolenic flax). Canadian Journal of Plant Science. 2007;**87**(2):297-299

[30] Gebauer SK, Psota TL, Harris WS, KrisEtherton PM. n-3 fatty acid dietary recommendations and food sources to achieve essentiality and cardiovascular

benefits. American Journal of Clinical Nutrition. 2006;**83**:1526S-1535S

[31] Goyal A, Sharma V, Upadhyay N, Gill SM. Flax and Flaxseed Oil: An Ancient Medicine & Modern Functional Food. Association of Food Scientists & Technologists (India); 2014. DOI: 10.1007/s13197-013-1247-9

[32] Liu F, Chang X, Hu X, Brennan CS, Guo X. Effect of thermal processing on phenolic profiles and antioxidant activities in Castanea mollissima. International Journal of Food Science and Technology. 2017;**52**(2):439-447. DOI: 10.1111/ijfs.13299